Gateways to Science

Neal J. Holmes
Department of Chemistry
Coordinator of Science Education
Central Missouri State University

John B. Leake
Department of Science Education
University of Missouri

Mary W. Shaw
Department of Education
Montgomery County, Maryland

Webster Division, McGraw-Hill Book Company
New York/St. Louis/Dallas/San Francisco/Auckland/Bogotá/Guatemala
Hamburg/Johannesburg/Lisbon/London/Madrid/Mexico/Montreal/New Delhi
Panama/Paris/San Juan/São Paulo/Singapore/Sydney/Tokyo/Toronto

Consultants and Reviewers

Alden Loomis   Montebello Unified School District, Montebello, California
Patricia F. Erikson   Ryan Road School, Northampton, Massachusetts
Dr. John P. Huntsberger   Science Education Center, The University of Texas at Austin, Austin, Texas
Dr. Charles J. LaRue   Montgomery County Schools, Rockville, Maryland
Miriam R. Buckland   K. T. Murphy School, Stamford, Connecticut
Joseph T. Piazza   Easton Area School District, Easton, Pennsylvania
James E. Williamson   Lamar Junior High School, Irving, Texas
Alice Kimbler Hankla   The Galloway School, Atlanta, Georgia
Saralee Semon   School Four, Fort Lee, New Jersey
Dr. Floyd D. Ploeger   Southwest Educational Development Laboratory, Austin, Texas

Editor in Chief: John F. Mongillo
Sponsoring Editor: Dennis Colella
Editorial Assistant: Helga Rauch

Manchester Publishing Unit
Project Leader: Neysa Chouteau
Editorial Management: Richard Paul
Editing and Styling: Robert Towns
Design and Art Supervision: E. Rohne Rudder
Production: Tom Goodwin
Editorial Assistant: Frieda Amiri

Editor: Barbara Branca
Associate Editor: Jane O'Connor
Photo Researcher: Randy Matusow
Cover Photo Research: Alan Forman
Text and Cover Design: Group Four, Inc.
Cover Photo: Jeff Rotman/Peter Arnold, Inc.
Layout and Art Production: Cover to Cover, Inc.
Artist: Roberta Collier
Contributing Writers: Wayne Leibel, Cynthia Kaufmann, Bertram Siegel

This book was set in 12 point Aster by Black Dot. The color separation was done by Black Dot.

ISBN 0-07-029914-5

Contents

Life science

Classifying living things 3

What is a group? 4
Grouping living things 5
Kingdom to species 6
Activity—Making a classification system 7
Naming organisms 8
Differences within a species 9

Fauna and flora 11
Activity—Plants in your neighborhood 11
Science in our world 13
Careers 13
Review the main ideas 15
Chapter review 16

Green plants 19

Green plants produce food 20
Activity—Does a plant need sunlight? 21
Making food 22
From sugar to starch 23
Activity—Test for starch 24
From sugar to fats and oils 25

From sugar to proteins 26
Plants and oxygen 26
Science in our world 28
Careers 28
Review the main ideas 29
Chapter review 30

Insects 33

What is an insect? 34
Meet the ants 36
Activity—Life cycle of an insect 38
The queen of the bee society 39
The busy bee 40

Learning about insects 42
Activity—Friends or strangers? 43
Science in our world 44
Careers 45
Review the main ideas 45
Chapter review 46

Birds 49

4

General characteristics of
  birds 50
**Activity—Make a bird
  feeder** 52
A most skillful hunter 53
Special characteristics of birds
  of prey 54
Activity—Birds of prey 56
Food chains 57

Vanishing birds of prey 60
Trouble in the food chain 61
Bringing back the peregrine 62
Science in our world 64
Careers 64
Review the main ideas 65
Chapter review 66

Mammals 69

5

What is a mammal? 70
Where mammals live 71
Three main groups 72
Activity—Animal names 75
Different kinds of "true"
  mammals 76

Activity—Animal communication 78
More "true" mammals 79
Science in our world 80
Careers 80
Review the main ideas 81
Chapter review 82

Bones, muscles, skin 85

6

Holding it all together 86
Activity—Name the bones 87
Living bones 87
Holding your bones together 90
Tying your joints together 91
Making your move 92
**Activity—Look at the
  drumstick** 94

Wrapping you up 95
Science in our world 97
Careers 98
Review the main ideas 99
Chapter review 100

The eyes and vision 103

7

Letting in light 104
**Activity—Watch your pupils
  change** 105

Images in the eyes 106
Focusing near and far 107

Color and the eye 108
Seeing what is not there 109
Activity—Optical illusion 110
Science in our world 112

Careers 112
Review the main ideas 113
Chapter review 114

Keeping fit 117

8

How to be your own coach 118
Time to sleep 119
**Activity—How do you spend
  your time?** 121
Standing tall 122
Weighing in 123
**Activity—Count the
  calories** 124

Feelings are fine! 125
Science in our world 126
Careers 126
Review the main ideas 127
Chapter review 128

What's in your food? 131

9

Then and now 132
Preserving foods 133
Coloring foods 135
Processing foods 136
Adding nutrients to food 138
**Activity—Experiment with
  additives** 139

Laws about additives 140
Activity—Reading food labels 141
Science in our world 142
Careers 142
Review the main ideas 143
Chapter review 144

Earth and space science

Minerals and rocks 149

10

What is a mineral? 150
Ores and gems 151
Identifying minerals 152
**Activity—Mineral streak
  test** 153
Rocks 155

**Activity—Identify rocks in
  buildings** 158
The rock cycle 158
Science in our world 160
Careers 160
Review the main ideas 161
Chapter review 162

Soil: a resource 165

11

Things that we value 166
Our food-producing lands 166
Soil conservation 168
Activity—Where does mud come from? 169
Protecting soil from erosion 170
Other conservation problems 172

Activity—Renewable resources 173
Science in our world 174
Careers 175
Review the main ideas 175
Chapter review 176

Conserving water 179

12

Two sources of water 180
The water cycle 181
Problems with water 182
City water supplies 184
How to clean water 185
Activity—Filter water 186
Is clear water always pure? 188

Conserving water 188
Activity—How well do you conserve water? 189
Science in our world 190
Careers 191
Review the main ideas 191
Chapter review 192

Weather watch 195

13

How air moves 196
Temperature 197
Activity—Comparing temperatures 198
Humidity 200
Air pressure 201
How clouds form 202
Cloud shapes 202

Gathering weather data 204
Reading weather maps 205
Activity—Using weather maps 207
Science in our world 208
Careers 209
Review the main ideas 209
Chapter review 210

Our solar system 213

14

A very special star 214
Our nearest neighbor 215
Activity—Phases of the moon 218
The planets 219
Other planets and their moons 222

Activity—A trip to the planets 223
The solar system 224
Science in our world 226
Careers 227
Review the main ideas 227
Chapter review 228

15 New sources of energy 231

Using geothermal energy 232
Garbage power 233
Energy from the sun 234
**Activity—Comparing
  temperatures** 235
Using wind energy 236
Energy from atoms 237
**Activity—Reading about
  nuclear energy** 239

Radiation 239
Science in our world 240
Careers 241
Review the main ideas 241
Chapter review 242

Physical science

16 Static electricity 247

Electricity at rest 248
**Activity—Making an electric
  charge** 248
Two kinds of charges 250
How objects are charged 250
Friction and static electricity 252
**Activity—Producing static
  electricity by friction** 252

Attract or repel 254
Loss of charges 255
Science in our world 256
Careers 256
Review the main ideas 257
Chapter review 258

17 Electricity and circuits 261

Energy and work 262
Conservation of energy 264
Electricity in circuits 264
Conductors and insulators 265
**Activity—Working with
  circuits** 266
Series circuits 267
**Activity—Working with a
  series circuit** 268
Parallel circuits 269

Two kinds of batteries 269
Electrical generators 271
Electricity and magnetism 272
Magnets with electricity 273
Science in our world 275
Careers 276
Review the main ideas 276
Chapter review 278

Sound 281

18

Stop and listen 282
How are sounds made? 282
How do we hear? 285
The speed of sound 286
Different qualities of sound 288
Activity—Recording sounds on tape 290

Sound as communication 291
Activity—Sounds and noise 294
Science in our world 295
Careers 296
Review the main ideas 297
Chapter review 298

Light and lenses 301

19

Light in motion 302
Activity—Studying refraction 303
Lenses 304
Refracting telescopes 305
Reflecting telescopes 306
Activity—Finding out about telescopes 307

Binoculars 308
Hand lenses and compound microscopes 310
Science in our world 312
Careers 313
Review the main ideas 313
Chapter review 314

Time and distance 317

20

Measuring distance 318
Average speed 318
Activity—Finding average speed 319
Picturing speed 320
Activity—Making a distance-versus-time graph 321

Being on time 322
Schedules 323
Science in our world 324
Careers 325
Review the main ideas 325
Chapter review 326

21 Matter 329

What is matter? 330
Three states of matter 331
Activity—The movement of molecules 332
Elements 333
Compounds 334

Activity—Classifying elements and compounds 335
Science in our world 336
Careers 336
Review the main ideas 337
Chapter review 338

22 Matter changes 341

Rusting 342
Activity—Making rust 342
Chemical changes 344
Activity—Chemical change 345
Physical changes 346

Science in our world 347
Careers 349
Review the main ideas 349
Chapter review 350

23 Communicating today 355

Electronic communications 356
Activity—Communications timeline 357
Communications satellites 358
How satellites transmit messages 359
Computers 361
Computers and printing 365

Home computers 367
Science in our world 368
Activity—How do libraries store information? 369
Careers 370
Review the main ideas 371
Chapter review 372

Glossary 374
Index 381

Life science

1. Classifying living things
2. Green plants
3. Insects
4. Birds
5. Mammals

Health and safety

6. Bones, muscles, skin
7. The eyes and vision
8. Keeping fit
9. What's in your food?

Classifying living things

In this chapter, you will find out that

■ An organism is a living thing.

■ The world has millions of kinds of organisms.

■ Scientists classify organisms into groups according to their characteristics.

■ Kingdoms are the largest groups of different organisms; species are the smallest groups of the same kind of organisms.

Before you begin

Did you know that there are more than two million different kinds of plants and animals in the world? There are so many kinds that nobody could ever learn all about each of them. How can scientists find and study so many different kinds of living things?

To make learning about plants and animals more manageable, scientists put them into groups according to their likenesses and differences. How would you group all the living things in your neighborhood? What would you do first? Would you have many groups? Would it be easier to work with a large group or with a small group?

What is a group?

Do you remember studying about groups and sets in math class? A group is made up of two or more things that are alike in some important way. Many things are **classified** (klăs′ə fīd′), or grouped. In the library, books are grouped. All of the books in each section of the library are grouped because they have something in common. In what other places are things grouped because they have something in common?

A flock of pigeons is a group. 1 How are pigeons alike? To what group do pigeons belong?

How do you know whether something is a member of a group? All members of a group have something in common. That is, they have one or more of the same **characteristics** (kăr′ĭk tə rĭs′tĭks), or features. For example, birds have feathers.

You can decide whether something is a member of a group. For example, a characteristic of pine trees is that they produce seeds in pine cones. 2 Pine trees also have long, needle-like leaves. Other kinds of trees produce seeds that are not in pine cones. These trees have broad, flat leaves. 3

classify:
To put things in groups according to their likenesses and differences.

characteristics:
The important or special features of a certain person or thing.

1

4

2   3

If pine trees were mixed up with other types of trees, what would be the easiest way to group them?

Grouping living things

Scientists also classify, or group. They classify **organisms** (ôr′gə nĭz′əmz), or living things. Scientists classify organisms to help identify them. Grouping also helps scientists see the relationships among the groups.

More than 2,000 years ago, a Greek philosopher (fĭ **los′**ə fər) named Aristotle classified living things into two large groups. These groups were plants and animals. Living things that moved were animals. Living things that did not move were plants.

Scientists tried many different ways to classify living things. But, the scientists always classified a living thing as a plant or an animal.

After the **microscope** (mī′krə skōp′) was invented, scientists discovered many small organisms. These organisms were classified into a third group. The name of this group is **protist** (prō′tĭst). Protists have several characteristics that are the same. The protists are microscopic. 4 These living things are neither plants nor animals. Today, some scientists use more than three major groupings.

organism:
A living thing.   4

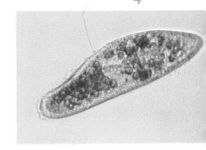

microscope:
A tool that lets us see things that are too small to see with the naked eye.

protist:
Microscopic organisms that are neither plants nor animals.

5

Check yourself

1. How do you decide whether something is a member of a particular group?
2. Are all living things members of a group?
3. How does grouping living things help scientists?
4. How did Aristotle separate animals from plants?
5. Name two groups that scientists have always used to classify living things.

Kingdom to species

Scientists use a special system to classify organisms. All organisms share characteristics with each other. These shared characteristics make organisms similar. Other characteristics make organisms different from each other. The characteristics of organisms are used to group living things.

Organisms are divided into groupings called **kingdoms**. You are going to study two kingdoms. One is the plant kingdom. The other is the animal kingdom. Today, some scientists use four, five, or six kingdoms.

Kingdoms are the largest grouping. They have the largest number of different kinds of organisms that share certain characteristics. For example, an ant and an elephant are both members of the animal kingdom. But, they are very different from

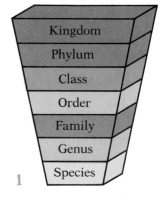

1

kingdom:
The largest group under which organisms are classified.

6

each other. For this reason, scientists further classify animals and plants into smaller and smaller groupings.

There are seven major groupings all together, starting with kingdom and ending with **species** (**spē′shēz′**). 1 Members of the same species have many, many characteristics in common. They can also reproduce the same kind of organisms. For example, all house cats are members of the same species.

species:
The smallest group into which organisms are divided.

Activity

Making a classification system

1. Obtain six small objects commonly found in the classroom or around your house. 2

2. Group the six objects into two large groups based on a characteristic common to each group.

3. On a piece of paper draw two large boxes. Write the common characteristic of a group above each box. Then group each of the six objects into the proper boxes.

4. Try to group the objects in each box into smaller and smaller groups until each group has only one object. Be sure to list a common characteristic for each group.

5. When you finish, write the name of each object and list its characteristics. If you were classifying organisms, what would be the name of the largest group?

2

As organisms are classified, they are grouped into smaller and smaller groupings. Fewer and fewer different kinds of organisms are placed into each grouping. The organisms in each group share more and more of the same characteristics.

Naming organisms

At first, naming the organisms was a problem because scientists come from all over the world and speak many different languages. A maple tree has a different name in English than it does in Italian or in Chinese. How would a scientist from one country know which plant a scientist from another country was talking about? In the 1700s, a plant scientist from Sweden, Carolus Linnaeus (**kă′**rə ləs lĭ **nē′**əs), solved the problem. 1

1

Linnaeus decided to give Latin names to the different species of plants and animals. Latin was a language used by scientists worldwide. Linnaeus gave each organism a two-part name. For example, the cat family is a large group. Each member of the cat family has two Latin names. A cat that you would have around the house as a pet is called *Felis domesticus*. 2 A tiger is called *Felis tigris*. 3 The Latin word *Felis* tells you that the animal is a member of the cat family. But there is a big difference between a tiger and a house cat! So the second word—for example, *tigris*—tells you to which species the animal belongs. *Tigris* is Latin for tiger.

2

3

✓ **Check yourself**

1. What is the grouping name that has the largest number of different kinds of animals?

2. In which grouping do living things have the most common characteristics?

3. Lions belong to the cat family. The word "leo" refers to a lion. What is the two-part name for a lion?

Differences within a species

All members of a species have many, many characteristics in common. But each member of a species may be different from all the others in some small way. Finding differences within a species takes careful study and observation.

How many differences would you expect to find in the fruit of the same species? Are all watermelons alike? 1 Are all squashes alike? How would you go about finding differences between one watermelon and another? You might study colors and markings. You might weigh each watermelon. You might measure the length and width of each. At first glance, different watermelons might look the same. But, after close study, you might find differences that would separate one from another.

1

Just as there can be differences between members of the same species, sometimes there are only slight differences between one species and another. Look at the animals in the photographs. 2 At first glance, you might say that they are alike. After some more observation, you might decide that they are different members of the same species. In fact, the zebras pictured are members of different species. Study the pictures carefully. Can you begin to find some differences you did not see at first? What characteristic can you find in one zebra that you cannot find in another zebra? A good scientist pays very close attention to tiny details.

2

Fauna and flora

You can make a list of all the animals or all the plants in your neighborhood. The group of all animals found in an area is called a **fauna**. The group of all plants found in an area is called a **flora**.

fauna:
All the animal life in an area.

flora:
All the plant life in an area.

Activity

Plants in your neighborhood

1. Take a tour of your neighborhood with a small group of friends.

2. Make a list of ten plants that you see. Draw a sketch of each plant. Make a list of its important characteristics. Write down its common name. If you do not know a plant's name, look it up or ask your teacher.

3. Choose three plants from your list. List any characteristics that they share. List any characteristics that are not shared.

If you are very ambitious, you may want to make a herbarium (hûr **bâr′**ē əm). 1   A herbarium is a collection of plants. In a herbarium, the plants have been carefully collected to include roots, stems, leaves, and flowers. Each plant is dried and mounted on a sheet of paper. A label is put on the sheet with the plant, telling the name of the plant and where it was found. The name of the collector and the date is also put on the sheet.

Check yourself

1. How do the members of a species differ from each other?

2. What is all the plant life in an area called?

3. What is all of the animal life in one area called?

4. What is a collection of dried plants mounted on paper called?

Science in our world

The reason that each species needs its own name is easy to see. When scientists speak of a species, they want to be sure that other people know just what species they are talking about. Having separate names for separate species makes this possible. As you know, Linnaeus used Latin to name the various species of plants and animals. Scientists from all countries knew Latin. Do you know why?

Many centuries ago, most people could not read or write. There were very few schools, and these schools were connected to churches. Latin was the language of the churches. So the schools taught their students in Latin. The ability to speak Latin was the sign of being educated.

In those days, the scientific studies were made at the schools run by the churches. Since Latin was the language of the schools, all scientific notes were written in Latin. Thus, the earliest names given to flora and fauna were Latin names. As time passed, the use of Latin died out. More and more schools were built. People began to be educated in their own languages. But the use of Latin names for species has continued.

Careers

Scientists who study plants and animals are called **biologists** (bī ŏl'ə jĭsts). "Bio" comes from the Greek word for "life." Biologists are people who study forms of life. Biologists who specialize in

studying flora are called botanists (**bŏt′n** ĭsts). 1
Biologists who specialize in studying fauna are
called zoologists (zō **ŏl′ə** jĭsts). 2

Biologists have to finish six to eight years of
college after high school. They like the outdoors,
and they like to study living things. Biologists
observe things very, very carefully. Often, they are
fascinated by the tiny characteristics that separate
one organism from another.

1

2

Review the main ideas

Organisms are living things. There are millions of organisms in the world. Scientists classify organisms into groups. The largest groups are called kingdoms. Two kingdoms are plants and animals.

Kingdoms are further divided into smaller groups. The group that scientists find most useful for study is the species, the smallest group.

All members of a species have many important characteristics in common. But they also differ from each other in small ways. Often the differences are so small that it takes an expert eye to see them. Members of one species also differ from members of another species.

Although different species may share a common name, each species also has its own scientific name. Scientists who speak different languages use the same scientific name when they are discussing a particular species.

Carefully observing the plants in your own neighborhood will help you to understand how scientists group living things. You will also see why classifying plants and animals makes it easier to learn about them.

Chapter review

Science vocabulary

Choose the items in column B that best match the terms in column A.

A	B
1. organism	a. the smallest groupings of organisms
2. classify	b. the special features of a person or thing
3. flora	c. a living thing
4. fauna	d. all the animals in a particular place
5. characteristics	e. to put things in groups
6. species	f. the largest groups of organisms
7. kingdoms	g. all the plants in a particular place

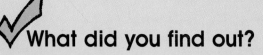 What did you find out?

Answer the following.

1. Biologists who study fauna are known as _____ .
2. The scientific name for a house cat is _____ .
3. Members of the same species have many of the same _____ .
4. A Swedish scientist named _____ set up a way to _____ organisms.
5. The roots, stems, leaves, and flowers of a plant are collected to make a _____ .
6. Biologists who study flora are called _____ .
7. Each plant and animal has (one/two) Latin names.
8. The characteristics of the members of one species (can/cannot) be different.

Think and write

1. What characteristics do all humans have?
2. Classify your personal belongings. Start with personal items and include everything you own.

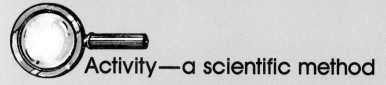

Activity—a scientific method

Purpose: to make a herbarium.

Materials needed

bricks or heavy weights	spoon	water
cellophane tape	construction (art) paper	newspapers

Procedure

1. Collect a plant from your neighborhood. Use the spoon to dig it out of the ground. Get the whole plant, including the roots. Wash it.
2. Sketch the plant, showing its shape and parts.
3. Put your plant between two pieces of newspaper. Fold a whole newspaper around the pieces that hold your plant.
4. Place heavy weights, such as bricks, on top of the newspaper.
5. When the plant is dry, tape it on a sheet of art paper. Label the sheet with the plant name and where it was found.

Observations

1. Does your plant look like any other plant in your neighborhood?
2. How is your plant like others in your neighborhood?

2

Green plants

In this chapter, you will find out that

- Green plants produce a kind of sugar by means of a process called photosynthesis.
- Photosynthesis can only take place in plants that contain chlorophyll.
- The extra sugar made by a green plant is changed by the plant into other nutrients.
- During photosynthesis, green plants use carbon dioxide and give off oxygen.

Before you begin

"Did you thank a green plant today?" Perhaps you have seen that question on a bumper sticker or poster. Why should people be thankful to green plants? Think about it.

Green plants make up an important part of our surroundings. We like green lawns and hillsides. We enjoy sitting in the shade of trees on hot, sunny days. But the beauty of green plants is not the most important reason to be thankful for them.

Biologists know that green plants make their own food. The plants use raw materials around them to do this. Green plants also release an important gas into the air. All animals depend on the food made by green plants and on the gas the plants release into the air. What would happen if there were no more green plants on Earth?

Green plants produce food

All living things need food in order to go on living. Plants take in nutrients from the soil and air to make food. Some animals eat plants. Others eat other animals. Some animals eat both plants and animals. But only green plants can make their own food. This is why green plants are called **producers**. How do they produce food?

Have you ever sat down on the grass when it was damp? What happened to your clothing? You most likely got grass stains on your clothes. The green stain was caused by a material called **chlorophyll** (**klôr′ə** fĭl). Chlorophyll is a pigment, or coloring material, found in green plants.

All green plants contain chlorophyll. The chlorophyll in green plants traps light energy from the sun. Then the plants use this light energy to change carbon dioxide and water into food.

The green plant gets water from the soil and the gas, carbon dioxide, from the air. It gets light energy from the sun. The process, or way by which these things are put together to make food, is called **photosynthesis** (fō′tō **sĭn′**thə sĭs). 1

"Photo" means light, and "synthesis" means to put together. So, photosynthesis simply means putting together light energy, water, and carbon dioxide to make food.

If a plant did not contain chlorophyll, photosynthesis could not take place. The plant would not be able to make food.

Remember, green plants need chlorophyll, water, carbon dioxide, and sunlight. What do you think happens when a green plant does not get sunlight? Try doing the following activity and see for yourself.

producer:
A living thing that can make its own food.

chlorophyll:
A coloring material that is found in green plants.

photosynthesis:
The way in which green plants make food.

Energy from the sun

Carbon dioxide

Water

1

Activity

Does a plant need sunlight?

1. You need two green plants. It is best if both plants are the same type.

2. Water both plants.

3. Place one plant on a sunny windowsill. Place the other plant in a dark closet.

4. When the plant on the windowsill needs water, water both plants.

5. Check both plants after one week. Compare them again after two week˜

6. Record what you see.

Both plants were watered. The plant in the closet got no light. It became droopy. That is because it could not make food. 2

2

21

Making food

Green plants are the primary, or most important, producers of food. The food produced by green plants during photosynthesis is a kind of sugar. Sugar is a **nutrient** (n$\overline{oo}$'trē ənt). A nutrient is the part of food that your body needs in order to grow or work. Some plants are grown mostly for their sugar. One such plant is sugar cane. 1   Another is

nutrient:
The part of food the body needs to help it grow or work.

22

the sugar beet plant. 2 We get molasses from sugar cane. Table sugar, powdered sugar, and brown sugar come from sugar cane or sugar beets.

From sugar to starch

Green plants produce more sugar than they can use. What happens to the extra sugar? Some is stored in plant roots and some in seeds and fruits.

3

4

When plants store sugar, they begin to change it into starch. Starch is also a nutrient. And, like sugar, starch is used by the body for energy. So, you can see that it is a very important part of our diet. We use the starch stored in wheat to make flour. 3 The flour is used to make bread and many other foods. 4 When we eat a banana, a potato, corn, or rice, we are eating starch. Potato plants store lots of starch in their underground stems. Banana plants store starch in their fruits.

23

Activity

Test for starch

1. You need iodine, bread, and a potato.
2. Have your teacher cut a piece of potato.
3. Place a few drops of iodine on the bread. Record what happens.
4. Place a few drops of iodine on the potato. Record what happens.
5. What is in both bread and potatoes?

Both bread and potatoes contain starch. When iodine combines with starch, it changes color. It turns blue-black.

Check yourself

1. What kind of food is produced during photosynthesis?
2. What two plants are grown mostly for their sugar?
3. Name a nutrient that plants produce from sugar.

From sugar to fats and oils

Two other nutrients made from sugar are fats and oils. Fats and oils are similar. But fats are solid and are produced mostly by animals. Oils are liquid and are produced mainly by plants. Our bodies use fats and oils for energy. Some plants are such good producers of oils that farmers raise them just for that purpose. Cottonseeds, corn, olives, coconuts, and sunflower seeds are all good sources of oil. 1

Plants Rich in Oil

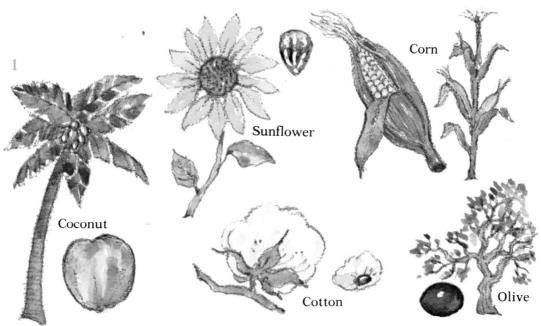

Coconut

Sunflower

Corn

Cotton

Olive

Most of the oils in a plant are stored in or near the seeds. Sometimes, plant oils are made into spreads for bread. These spreads are called margarines (**mär′**jər ĭnz). Margarine is different from butter. Butter is made from animal fats found in milk. Margarine is made from plant oils.

From sugar to proteins

Green plants also change sugar into **proteins (prō′tēnz)**. Proteins are nutrients that the body uses to build muscles, nerves, and other body parts.

Lentils, soybeans, and peanuts are just a few plants that humans eat for protein. Other protein-rich plants, such as alfalfa, are fed to cattle and chickens. The cattle and chickens change the plant protein into muscle.

When we eat beef and poultry, we are eating protein. It is animal protein instead of plant protein. But it started out as plant protein.

protein:
Nutrient used by the body to build muscles, nerves, and other body parts.

Plants and oxygen

Has anyone ever said to you, "Take a big breath of fresh plants"? Probably not. But did you know that, if it were not for plants, human beings most probably would not be able to breathe? Why is this true?

Oxygen (ŏk′sĭ jən) is a gas found in the air. People need oxygen for life. We breathe in air. Oxygen moves from our lungs into our blood. As the blood travels through the body, the oxygen is taken out and used by the body. When we exhale, or breathe out, we give off a gas called **carbon dioxide (kär′ bən dī ŏk′sīd′)**. Our bodies cannot make use of carbon dioxide.

But plants can use carbon dioxide. Plants take carbon dioxide out of the air and use it during photosynthesis. Besides making food during photosynthesis, plants also produce oxygen. 1 Some of the oxygen is kept by the plants for their own needs. But most of it is given off into the air.

oxygen:
A gas found in the air and needed for life.

carbon dioxide:
A gas found in the air and used by plants during photosynthesis.

26

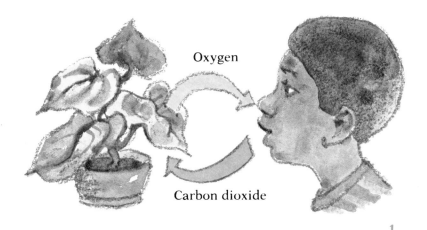

Oxygen

Carbon dioxide

If plants stopped producing oxygen, we would have a hard time breathing. People and animals cannot make oxygen. Without green plants, the oxygen in the air would probably be used up.

Thus, you can see that photosynthesis is a very important process. Without it, we might have no food to eat and no oxygen to breathe.

Check yourself

1. Where are oils stored in a plant?
2. What is the difference between butter and margarine?
3. Why are plants important to people?
4. Name a gas that plants use to make food.
5. Why is oxygen important?

Science in our world

Many plants are used as food by people. In season, fresh fruits and vegetables are sold at markets. 1 By freezing or preserving food, we can enjoy eating plant foods long after the growing season is over.

Some of the food produced by plants is fed to animals. The meat from these animals is eaten by people. Sheep, goats, pigs, cattle, and chickens provide much of the meat people eat. People drink the milk from cattle and goats. This milk also is used to make cheese. Can you name other plant foods we use?

Plants must be protected so that the world will always have a supply of food. The amount of land available for growing crops is decreasing. It is being covered by streets and buildings. Also, when water pollution increases, many water plants cannot grow. These plants give off the oxygen that fish need to live. Fish is another food source for people.

1

Careers

2

Each year, more and more people are born. The world's population increases. More food is needed to feed more people. But there is only a certain amount of land that can be used for raising food. This means that more crops have to be grown on the same amount of land. In other words, crop yields must be increased.

Agronomists (ə grŏn'ə mĭsts) are scientists who help farmers improve their crop yields. 2 Agron-

omists develop new plants and new growing methods. They work to control plant diseases and pests that harm plants. They study soils. Agronomists help farmers with ways to decrease soil erosion.

Review the main ideas

Green plants are called "producers" because, directly or indirectly, they produce most of the food we eat. Green plants also produce the oxygen we breathe.

During a process called photosynthesis, the chlorophyll in green plants traps light energy from the sun. The plants use this energy to change water and carbon dioxide into sugar. Oxygen, a gas found in the air, is also produced during photosynthesis. Green plants change the extra sugar they make into starches, proteins, fats, and oils. All of these are nutrients needed by the human body. In addition to eating plant food directly, humans feed some plants to animals. In turn, humans eat the animals.

Humans and animals breathe in the oxygen produced by plants. Humans and animals breathe out carbon dioxide. The plants take in carbon dioxide and use it in photosynthesis. Thus, a cycle is kept going. People breathe in oxygen and give off carbon dioxide. Plants take in carbon dioxide and give off oxygen.

There is a real need to produce greater amounts of food in smaller amounts of space. Because of this need, a new science was formed. It is called agronomy. It is the work of the agronomist to help improve crop yields. Agronomists work to grow greater and greater amounts of food plants on smaller and smaller amounts of land.

Chapter review

Science vocabulary

Choose the items in column B that best match the terms in column A.

A

1. photosynthesis
2. oxygen
3. nutrient
4. producer
5. chlorophyll
6. protein
7. carbon dioxide

B

a. a living thing that can make its own food

b. a nutrient used to build muscles and nerves

c. the process, or way, in which green plants make food

d. a gas used by plants during photosynthesis

e. a material the body needs to help it grow

f. a gas found in the air and needed for life

g. a pigment that is found in green plants

What did you find out?

Answer the following.

1. To carry on photosynthesis, a plant must contain _____ .

2. Three things green plants use to make food are _____ , carbon dioxide, and light energy.
   (a) sugar   (b) nutrients   (c) water

3. Green plants use chlorophyll to trap
   (a) light energy   (b) gases in the air   (c) green dyes

4. A green plant produces _____ sugar for its own needs.
   (a) too much   (b) just enough   (c) too little

5. When we eat beef, we are eating _____ protein.

6. During photosynthesis, green plants give off _____ .

Think and write

1. Explain why you would thank a green plant.
2. Describe what happens during photosynthesis.
3. Make a chart listing the differences between plants and animals. Your list should include those differences found in the chapter. Give some examples of plants and animals. Illustrate your chart.

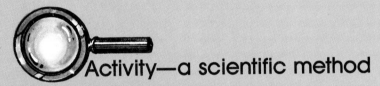

Activity—a scientific method

Purpose: to test a plant for oil.

Materials needed

peanut (shelled)   plastic sandwich bag   sheet of paper   pencil

Procedure

1. Put the peanut in the plastic sandwich bag.
2. Fold the paper in half. Put the bag between the two halves of paper and put everything on the floor. Gently crush it.
3. When the peanut is in fine pieces, put the bag on a flat surface and roll a pencil over it until the peanut is soft and creamy. You have made peanut butter. Taste some. Put the rest of the peanut butter on a sheet of paper. Let it stand for a while.

Observations

1. Describe what happened to the paper with the peanut butter on it.
2. Explain why it happened.

3

Insects

In this chapter, you will find out that

- Insects have characteristics that make them part of a group.
- Insect colonies are found all over the world.
- Ants and bees live in organized colonies.
- The members of a society do certain jobs.

Before you begin

Do you know how many insects there are? There are more than 300 million insects for every person living on the Earth.

Many insects help human beings. Bees, for example, provide us with honey. They also help flowering plants reproduce. Some insects harm human beings. Harmful insects destroy our crops. Others carry germs that spread diseases.

Many insects do interesting things. Have ants ever found their way to your picnic table? You probably noticed that where you see one ant, there are many more close by. That is because ants usually live together in large groups.

Ants are not the only insects that live in large groups. Bees do, too. In these groups, every member has a certain job that helps the whole group. Your community is a group. Can you think of jobs in your community that help the whole group?

What is an insect?

Did you know that there are more than 800,000 different kinds of insects in the world? 1 And scientists discover more than 8,000 new kinds of insects every year.

1

All insects have certain characteristics which make them part of the same group. Insects are **invertebrates** (ĭn **vŭr′**tə brĭts). Invertebrates are animals that do not have a backbone. Instead of bones, insects have a hard outside skeleton. It protects their bodies. Then how does an insect move? If you look closely at an insect, there are small joints in its legs. These joints let an insect move.

invertebrate:
Animal without a
backbone.

Insects have three body parts. 2 The first is the **head**. The middle part is called the **thorax** (thôr′ăks′). The end part is called the **abdomen** (ăb′də mən). On their heads, they have two **antennae** (ăn tĕn′ē). Have you seen the antennae of an insect? What do you think they are used for? Insects smell, taste, and touch with their antennae.

All insects have 6 legs. And, many insects have wings. How many legs does a spider have? 3 Is a spider an insect?

thorax:
Middle body part of an insect.

abdomen:
The end body part of an insect.

antennae:
Parts on an insect's head used for taste, smell, and touch.

2

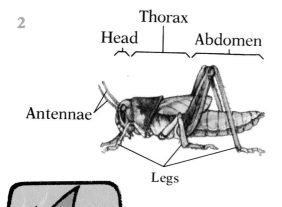

Thorax

Head | Abdomen

Antennae

Legs

3

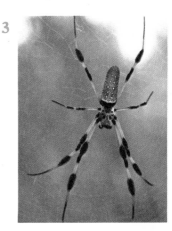

✓ Check yourself

1. What is an animal without a backbone called?

2. How many kinds of insects are there?

3. Where is an insect's skeleton?

4. Name the three body parts of an insect. On which body part are the antennae?

5. What do insects use to smell, taste, and touch?

6. Do all insects have six legs?

Meet the ants

Have you ever lifted a rock and seen ants? You probably uncovered a whole **colony** of ants. Ants are among the most common insects in the world. Colonies of ants live on all the continents except one—Antarctica.

Ant colonies are well organized groups, or **societies** (sə **sī'ĭ** tēz). Insects that live in societies are called **social insects**. Each member of the ant colony has a special job to do. The most important member is the queen ant. 1   The queen ant's work is to lay eggs to produce new ants. She is larger than the other ants. She also has wings. In this way, she can fly to a new place, lay the eggs, and start a new colony.

Most of the members of the ant colony are workers. The worker ants bring food into the colony. Worker ants are strong. A worker ant can carry an object many times its own weight. Imagine yourself picking up something as heavy as a car! Worker ants are also hungry. When they find something good to eat, they try very hard to bring it to the colony. Even if the food is much too big to move easily, the worker ants do not give up. They work together and keep trying.

Other worker ants raise the young ants. An ant goes through several changes from the time it first appears as an egg to the time it is fully grown. During these stages, adult worker ants care for the

colony:
Community of insects.

society:
An organized group of living things.

social insect:
Insect that lives in a society.

1

Queen ant

new ant. First, the queen ant lays the eggs. Then, the egg hatches into a **larva** (lär′və). A larva looks like a tiny worm. Then, the larva becomes a **pupa** (pyoo′pə). During the pupa stage, the ant-to-be rests in a covering called a **cocoon** (kə koon′). Finally, the ant is fully formed and becomes an **adult** (ə dŭlt′). 2

larva:
The first stage of insect life.

pupa:
The inactive stage in the life of an insect.

cocoon:
A protective covering made by some insects.

adult:
A fully formed insect.

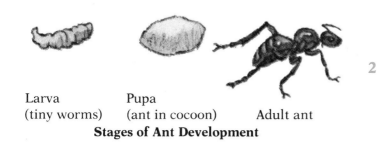

2

Larva   Pupa
(tiny worms)   (ant in cocoon)   Adult ant
        Stages of Ant Development

Another job of worker ants is to protect the ant colony from strangers. The worker ants fight ants from other colonies who try to move in. Some ant colonies have soldier ants. Their job is to help the worker ants protect the colony.

There are many different kinds of ants. Different kinds of ants do different things. Leaf-cutter ants grow their own food crops. 3   These ants cut off leaves from plants. Then, they use the leaves for "soil" on which to grow a fungus (fŭng′gəs). Fungus is similar to the mold that grows on stale bread. The fungus raised by "leaf-cutting" ants is the colony's source of food.

Do you think it is strange that ants grow food? Here is something even stranger. Some ants milk their own "cows." A certain type of ant colony lives with insects called aphids (ā′fĭdz). 4   The aphids eat lots of plant juices and become fat. Then, the ants "milk" the aphids to get the juices.

3

4

Activity

Life cycle of an insect

1. You will need a large glass jar with a lid, some oatmeal, a slice of potato, and shredded newspaper.

2. Cover the entire bottom of the jar with oatmeal. Place the slice of potato on top of the oatmeal. Add some mealworms to the jar.

2. Cover the oatmeal, potato, and mealworms with about 3 inches of shredded paper. Place the jar in a warm place that does not get too much light.

3. Keep a record each day of any changes you can see in the mealworms. Draw some pictures to show the changes.

Most insects go through a life cycle that has four parts. The cycle begins with the egg. A larva hatches from the egg. Did you know that a caterpillar is an insect larva? The mealworm is another kind of larva. It looks like a tiny worm. It crawls about and eats. After a few days, each mealworm larva becomes a pupa. How is the pupa different from the larva? The pupa does not move or eat. Each pupa will become an adult in about 3–10 days. It may surprise you that the adult is a beetle.

✓ Check yourself

1. Define "colony."
2. Why are ants called social insects?
3. What is the work of the queen ant?
4. List two things that are done by worker ants.
5. What is the first stage of insect life called?

The queen of the bee society

Like ants, most bees also live in colonies. The queen bee is as important to her colony as the queen ant is to hers. 1

The main job of the queen bee is to lay eggs. She lays between 1500 and 2000 eggs per day. She is attended by workers all the time. The workers groom her, feed her, and take care of the eggs, so that the queen can carry out her duties.

The busy bee

A worker bee lives for only about five weeks during the summer. Summer is the season when there is plenty of work for bees. The first two days of a worker bee's adulthood are spent cleaning the **hive** (hīv). A hive is the place where bees live. 1

On the third day, the worker bee feeds the older larvae. After the fifth day, the worker starts feeding the younger larvae. The worker is busy because each larva is fed about 1300 times a day.

Between the ages of 16 and 21 days, adult worker bees become field bees. Field bees look for flowers. 2 They gather nectar and pollen from the flowers and bring them to the hive. When a field bee returns to the hive, it can "tell" the other bees where the nectar was found. The bee does a special dance to show the others the way to the flowers.

A field bee may visit more than 2500 flowers in a day. In the hive, other worker bees make honey from nectar, using special glands in their bodies. It takes about 74,000 trips to the field to make one kilogram (a little over two pounds) of honey. For that much honey, the field bees must fly thousands of miles.

Where do you find beehives? The photograph shows one in a tree. Sometimes, they are in the

hive:
The home in which bees live.

1

2

ground or on buildings. People who keep bees for honey set up wooden boxes for the bees to live in.

No matter what the outside temperature, the temperature in the hive always stays the same. When it is cool out, the bees beat their wings inside the hive to produce heat. When the hive gets too warm, workers stand at the entrance and fan cool air in with their wings.

Sometimes, a small animal will come into a hive by mistake. What a deadly mistake! Bees protect themselves with stingers and will sting the intruder. 3 If the animal dies inside the hive, the bees will try to get it out. If the animal is too big for the bees to move, they produce a kind of wax to

Honeybee

Stinger

3

cover the dead animal. The wax forms a "mummy" around the animal. It keeps it from rotting and ruining the hive.

There may be as many as 60,000 bees in a colony during the summer. In the fall, many of the bees die. A few live through the winter to start the hive the next spring. In the hive, the bees work totally in the dark. So, the next time you eat honey, think of all the work the bees did.

You have been reading about some of the amazing activities of ants and bees. Can you imagine what it would be like to be one of the social insects?

Check yourself

1. Describe two jobs that are done by worker bees.

2. What does a field bee do that shows other bees the way to the flowers?

3. What do bees do when the hive gets too cool?

4. Describe how bees protect the hive when an intruder enters.

Learning about insects

One scientist interested in insects was Sir John Lubbock of England. He studied ants for more than 10 years. Sir John learned that ants can tell members of their own colony from members of other colonies. They do this by rubbing the antennae (ăn tĕn′ē) on their heads. This is the way they communicate with one another.

Sir John wondered how ants from the colony, "friends," and ants from different colonies, "strangers," would be received if they were in a drunken or intoxicated condition. He did an experiment. On November 20, 1879, Sir John Lubbock took the following notes:

The sober ants seemed puzzled by the intoxicated ants. They picked them up. They carried them about for a time aimlessly. At 11:00, I began to observe how sober ants received "friends" and "strangers."

At 11:30, a friend was carried to the nest.

At 11:50, a stranger was dropped into the water.

At 12:30, a stranger was dropped into the water.

At 12:31, a friend was dropped into the water.

At 1:10, a stranger was dropped into the water.

At 1:18, a stranger was dropped into the water.

At 1:27, a stranger was dropped into the water.

At 1:30, a friend (partly recovered) was taken to the nest.

At 2:30, a friend was picked up and carried about till 2:55. It was then taken to the nest. At the door the bearer met two other ants. The two ants seized the intoxicated one and carried it off. After a time, it was dropped into the water.

At 3:35, a friend was carried to the nest.

Activity

Friends or strangers?

Let us see if we can draw some conclusions from Sir John's careful observations.

1. Were strangers taken to the nest? What can you conclude about ants taken to the nest?

2. Were there more strangers or friends dropped into the water? What can you conclude about how the ants treated strangers?

3. Why do you think one of the friends carried to the nest was dropped into the water?

4. How many friends were just dropped into the water? Why was this done?

✓ Check yourself

1. What are ants doing when they rub the antennae on their heads?

2. Describe the experiment that Sir John Lubbock did to find out how ants communicate.

Science in our world

Ants usually eat dead insects, dead animals, and some kinds of plants. In the photograph, one red ant is giving food to the other. 1 Some ants, such as carpenter ants, tunnel through wood and weaken it. This can happen to the wooden floors in some houses. Termites, which are close relatives of ants, ruin wooden structures by eating wood.

44

Careers

For hundreds of years, people have used bees to supply honey and wax. Bees also are used to pollinate flowers for good seed production. Bee-keeping is a large industry in the United States and in the rest of the world. Experienced **beekeepers** know what bees need in order to live. 2 Beekeepers know how bees communicate with each other. They understand how to handle bees without getting stung. Would you like to earn a living working with insects? Then you may want to keep bees.

2

Review the main ideas

Insects are invertebrates. Their bodies are divided into three parts. They have six legs and many have wings. Ants and bees are social insects.

Every ant colony has a queen ant. It is her job to lay the eggs so that there will be new ants. The worker ants do many tasks. They bring food to the colony and care for the young ants. They also protect the colony from strangers.

The queen bee is the most important bee in the hive. She controls the bee colony. The queen also lays eggs for the colony.

Worker bees have many different jobs. They groom the queen, feed her, and take care of the eggs. They also feed the larvae and clean the hive. Sometimes, they stand at the door of the hive and fan the air. When the workers become field bees, they look for flowers to gather pollen and nectar. They bring them to the hive, where honey is made from them. Our knowledge about insects comes from the careful work of many scientists.

Chapter review

Science vocabulary

Choose the items in column B that best match the terms in column A.

A

B

1. society
2. colonies
3. larva
4. pupa
5. thorax
6. adult
7. antennae

a. middle part of an insect's body
b. an organized group of individuals
c. insect communities
d. hatches from an insect egg
e. follows the larva stage
f. insect parts used for taste, smell, and feel
g. final stage in the life cycle

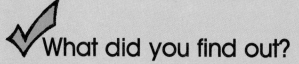

What did you find out?

Answer the following.

1. Two kinds of social insects are _____ and _____ .
2. Colonies of insects are found only in Africa. (True/False)
3. In a bee colony, the _____ take care of the young.
4. Field bees collect nectar so worker bees can make _____ .
5. Harvester and leaf-cutter are two kinds of _____ .

Think and write

1. What is a social insect? Give an example of a social insect and describe how it lives. Choose an example not given in this chapter.

2. Compare the job of a queen ant with that of a queen bee.

3. Imagine that you are a scientist. Describe the steps you would take to learn about a newly discovered or an unusual insect.

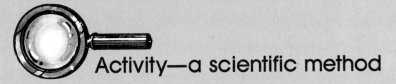

Activity—a scientific method

Purpose: to make an ant colony.

Materials needed

ant farm   egg white   honey   rubber band   melted butter
trowel   soil   water   piece of cloth   pop-bottle lid

Procedure

1. Put soil in the ant farm. Add a little water.

2. Find an ant colony. Dig up the colony with a trowel. Place the ants on the cloth. Try to include a queen.

3. Tie the ends of the cloth with the rubber band. Add the ants to your farm.

4. Make ant food this way: Mix 1 part honey, 2 parts water, and 1 part egg white. Add 1 part melted butter.

5. Put water in the pop-bottle lid. Place it into the farm. Put some ant food in one corner of the farm. Record your observations.

Observations

1. What is the first thing the ants do? Do all ants do the same thing?

2. Which ant works the least?

Birds

In this chapter, you will find out that

■ All birds share certain characteristics.

■ Every bird is adapted to the environment in which it lives.

■ Birds of prey hunt and eat other animals.

■ Animals at the end of long food chains are most threatened by chemical pollution.

Before you begin

The peregrine falcon in the photograph lives in the Sierra Nevada Mountains in western United States. Penguins make their homes in the cold climate of Antarctica. Robins and cardinals are among those birds that make their homes in the country. Pigeons can be found in cities and in the country. Seagulls, swans, and ducks live near water.

As you can see, different birds make their homes in different places. But no matter where a bird lives or what it looks like, all birds share certain characteristics. Every bird is adapted, or suited, to the environment in which it lives.

General characteristics of birds

Although birds come in different colors, sizes, and shapes, they all share certain important features, or characteristics. All birds begin life by hatching from, or coming out of, an egg that is covered with a hard shell. Birds lay many different kinds of eggs. 1

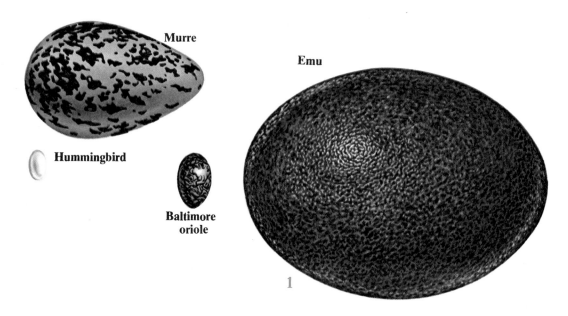

Murre

Emu

Hummingbird

Baltimore oriole

1

Each bird's body is covered with feathers. 2 The feathers grow in rows from the bird's skin. This characteristic of birds is very important. Feathers help to control the bird's body temperature and are necessary for the bird to fly.

All birds have four limbs. 2 One pair of limbs is the legs. They are used for hopping, walking, swimming, or running. The other pair of limbs is the wings. Some birds, such as Canada geese, have very powerful wings. They spend much of their life in flight. Birds with small wings are usually poor fliers. For example, chickens do not fly well.

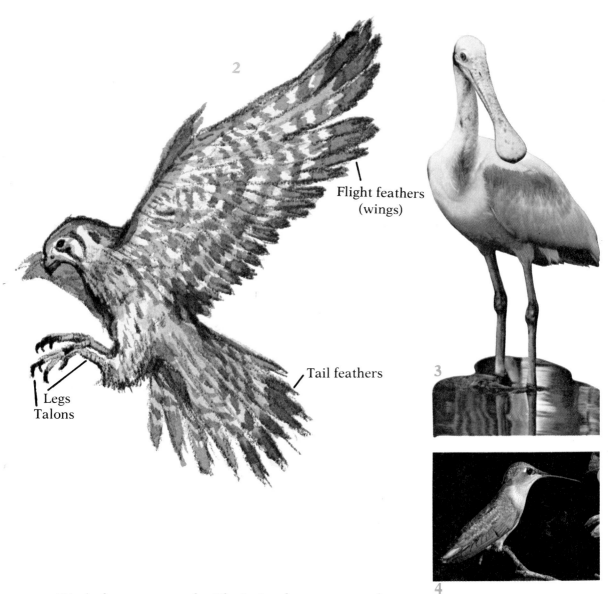

Flight feathers
(wings)

Tail feathers

Legs
Talons

2

3

4

Birds have no teeth. Their jawbones extend out to form bills, or beaks. The shape of a bird's bill helps the bird eat certain foods. For example, spoonbills use their bills to spoon fish and insects from the water. 3 The hummingbird has a very long, thin bill. 4 It is just the right size and shape for taking the sweet juices from flowers.

Each bird has a special combination of feathers, bill, and limbs. Because of this combination, a bird is **adapted** (ə dăpt'ĭd) to the environment, or place in which it lives. If you study a bird carefully, you can see how it is adapted to its environment.

Many types of birds are difficult to see close up. They avoid people. You may get a closer look at them by making a bird feeder.

adapted:
Suited to an
environment.

Activity

Make a bird feeder

1. You need a clean gallon plastic container, scissors, a jar-size lid, string, and a twig about 45 centimeters (1½ feet) long.

2. Using the scissors, carefully punch a hole in the front of the bottle, near the bottom. It should be big enough for the twig to go through. Now punch a hole on the other side of the bottle, opposite the first hole.

3. Push the twig through the holes. The ends that stick out will be a perch for birds.

4. Use the lid to trace a circle just above the perch. Carefully, cut out the circle.

5. Place birdseed and suet into the feeder. Then, use the string to hang the feeder from a tree.

6. Keep a record of birds that come to the feeder. Include information such as the kinds of birds that come, the time of day that they come, and their features (color, etc.).

Your bird feeder will attract small birds, especially in the autumn and winter. Once you have started to feed birds in the winter, you must continue until spring. Otherwise, the birds that have become dependent upon you for food may starve.

✓ Check yourself

1. List three characteristics that are shared by all birds.
2. Describe two things that are done by a bird's feathers.
3. What does the shape of a bird's bill do for a bird?
4. How does a spoonbill use its bill?
5. Define "adapted."

A most skillful hunter

The peregrine falcon is a very skillful hunter. Peregrine falcon chicks are adapted to a life of hunting. At birth, each chick only weighs about one ounce. 1 Its thin legs cannot even hold up its tiny body. But soon this will change. These weak chicks will grow up to be among the most skillful and powerful of hunters.

A peregrine hunts almost any bird that it can strike in midair. It hunts its **prey** (prā) from great heights. When ready to attack, the falcon folds its

1

prey:
An animal hunted by another animal for food.

53

long, pointed wings back along the sides of its body. From high in the sky, it falls like a rock, reaching a diving speed of as much as 290 kilometers (about 181 miles) per hour! With the force of this great speed, it can actually knock birds larger than itself out of the sky. The prey falls through the air and is caught in the falcon's **talons** (tăl′ənz) before it reaches the Earth. 1

Peregrine falcons feed on ducks, sandpipers, and gulls. They hunt over marshes and beaches. Dropping from above, peregrines may use their heavy yellow feet to strike their prey. Each yellow foot has four toes—three in front and one behind. The peregrine can use the curved talon on its back toe to strike its prey.

Special characteristics of birds of prey

Scientists call birds like the peregrine falcon "birds of prey." A bird of prey hunts and eats other animals. Eagles, vultures, and ospreys (ŏs′prēz) are also birds of prey. Like the peregrine falcon, these birds hunt during the day. Some other birds of prey, such as the owl, hunt at night.

How are birds of prey adapted to a life of hunting? Let us take a closer look at some. This red-shouldered hawk was flying high above the ground.

talon:
A long, sharp claw.
Birds of prey have a
talon on the end of
each toe.

1 Talons

2

But it had no trouble seeing a fieldmouse in the grass below. 2   The eyes of this bird can move rapidly. Also, the hawk's eyes are very large for the size of its body.

Because of the size and structure of their eyes, birds of prey can see much better than humans. This golden eagle can see eight times better than an average human. 3   How would you describe the eagle's beak? The upper bill is hooked. This is another characteristic common to all birds of prey.

3

Perhaps the most unusual characteristics of birds of prey are their feet. Look at the feet of the osprey. 4   When hunting for a fish dinner, the osprey uses the talons on its strong feet to seize and hold the fish. Excellent eyesight, strong feet, and a hooked beak adapt a bird of prey to its life as a hunter. The peregrine falcon and the other birds mentioned here are just some of the many birds of prey found all over the world.

4

Activity

Birds of prey

1. Find out the names of ten birds of prey.
2. Try to find pictures of them.
3. Make a chart. Include the name of the bird, where it lives, how it lives, special features, its flying habits, and how it raises its young.

Some of you may have done research on owls, hawks, eagles, kingfishers, and ospreys for this activity. These are only a few of the birds of prey.

If you are fascinated by birds, you should study their **migration** (mī **grā′**shən) habits. Migration is the movement of certain birds from one place to another at certain times of the year. Birds may migrate short or long distances. You may even live in a place over which migrating birds fly. 1

migration:
Seasonal movement
of birds from one
place to another.

Migratory Routes of Peregrine Falcon

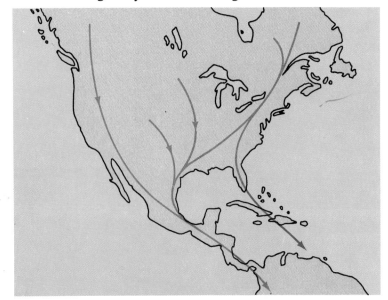

1

Check yourself

1. Define "prey."
2. Describe how the peregrine falcon hunts.
3. What is the job of talons?
4. Name three birds of prey.
5. List three characteristics of birds of prey.
6. What is the seasonal movement of birds from one place to another called?

Food chains

Animals get energy from the foods they eat. Different animals eat different kinds of foods. But no animal can make its own food. As you know, only green plants can use the sun's energy to make their own food. For this reason, green plants are called producers (prə $\overline{\text{doo}}$′sərz). Animals are called consumers (kən $\overline{\text{soo}}$′mərz) because they get their food by feeding on producers or on other animals. Some consumers, such as sparrows and ducks, are called **primary** (prī′měr′ē), or first, **consumers.** This is because they eat the producers directly. For example, ducks eat water plants.

Some animals eat other animals. A snake may eat a frog, a hawk may eat the snake, and so on. These consumers are known as **secondary** (sěk′ən děr′ē) **consumers.** The peregrine falcon is a secon-

**primary consumer:
Animal that eats
green plants
(producers) directly.**

**secondary consumer:
Animal that eats other
animals.**

57

dary consumer. Falcons eat primary consumers, such as ducks.

You can use a **food chain** to describe how energy in the form of food is passed from one living thing to another, or from producers to consumers. Sometimes, scientists draw a diagram with arrows to show a food chain. For example, ducks are one kind of food for falcons. And ducks feed on water plants. One food chain for the peregrine looks like this:

water plants ⟶ ducks ⟶ peregrine falcons

food chain:
A way to describe how energy, in the form of food, is passed from one living thing to another.

The drawing shows that water plants are eaten by ducks and ducks are eaten by peregrine falcons. 1

Peregrine falcons eat many other kinds of birds, too. These birds are part of other food chains. The food chains may look something like this. 2

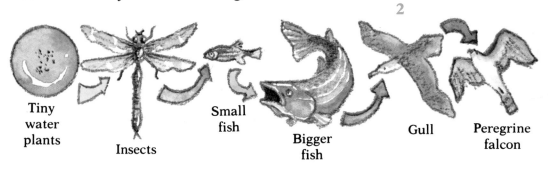

Tiny water plants

Insects

Small fish

Bigger fish

Gull

Peregrine falcon

As you can see, all food chains begin with green plants. The peregrine falcon is like other consumers. No matter what it eats, a food chain for the peregrine always begins with the producers, green plants.

Although birds of prey are parts of different food chains, they all occupy the same place in each chain. They are found at the end of it.

Like its fellow birds of prey, the peregrine falcon is at the end of a long food chain. In 1950, something strange was observed. Several kinds of birds of prey were disappearing. There was a decrease in the numbers of hawks, ospreys, and falcons in the United States. These birds were vanishing from parts of Europe and Russia, too. At about the same time, many people were starting to use DDT and other **pesticides** (pĕs′tĭ sīdz′). Pesticides are chemicals used to kill pests. Was there any relationship between the disappearance of the birds of prey and the growing use of pesticides? Scientists began to search for the answer.

pesticide: Chemical used to kill pests such as insects and rodents.

✓ Check yourself

1. Why are green plants called producers?
2. What kind of consumer is a peregrine falcon?
3. What is used to show how energy, in the form of food, is passed from one living thing to another?
4. What is the job of pesticides?

Vanishing birds of prey

In the spring of 1980, four chicks hatched in the marshes along the coast of New Jersey. 1 To wildlife biologists, this was a very important happening. The chicks were peregrine falcons. And they were the first of their kind to be born east of the Rocky Mountains in more than twenty years.

1

In another part of the country, high on a ledge 2000 feet above a valley, two chicks pecked their ways out of speckled, creamy pink eggs. Their nest is the only nest for peregrine falcons in all of the Sierra Nevada Mountains. And they were the first peregrine chicks to be born on the west coast of the United States in almost thirty years!

Why are there so few peregrine falcons today? What caused these wonderful birds to nearly vanish from the sky? Will they come back?

It took scientists more than twenty years to find the answer to the mystery of the vanishing birds of prey. They looked for clues in the nests of the peregrine falcons. There, they found broken eggs. Something was causing the eggs to break before the chicks could hatch. Then, scientists studied the eggshells. They discovered that the shells were very thin. You know that a parent bird sits on the eggs in the nest. Because the shells were so thin, the eggs were breaking. The peregrine falcons were dying off because no chicks were hatching from the thin-shelled eggs. But why were the eggshells of the peregrine falcons becoming so thin?

Trouble in the food chain

Scientists began to study the food chain to find the answer. They found DDT in the food chain and discovered that the amount of DDT increased with each link in the food chain. Let us see how this might happen. Pesticides are sprayed on crops. Rain washes the DDT off the land into the rivers and lakes. The tiny green plants that live in the water take in DDT. Pesticides are now part of the living food chain. 2   Tiny fish feed on the water plants. Pesticides become part of the bodies of these tiny fish. What do you suppose happens next?

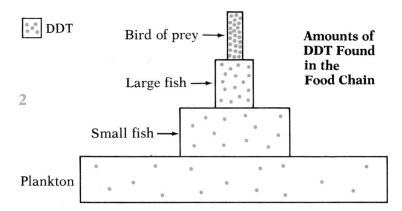

DDT

Bird of prey →

Large fish →

2

Small fish →

Plankton

Amounts of DDT Found in the Food Chain

An even greater amount of the pesticide collects in the bodies of birds that eat the fish. The peregrine falcon feeds on these small birds. As a result, the peregrine gets large amounts of pesticide in its body. The pesticide hurts the bird's ability to form eggshells. As a result, the peregrine falcon was dying off. It was in danger of becoming extinct.

In 1972, the use of DDT was outlawed in the United States. Since then, the use of other dangerous pesticides has been cut down. But, for the peregrine falcons, time was running out. Could these graceful hunters of the sky be saved from extinction?

Bringing back the peregrine

A rescue program was begun to bring back the peregrine falcon. As you have learned, peregrines usually nest and raise their young on the ledges of steep cliffs. Scientists had to locate the nests, climb steep cliffs, and carefully remove the fragile eggs. And this was only the beginning of the rescue program.

Dr. Tom Cade is a professor of **ornithology** (ôr′nə thŏl′ə jē) at Cornell University in Ithaca, New York. Dr. Cade was in charge of the peregrine falcon rescue program at Cornell. Peregrine eggs were placed in special incubators in a laboratory at the university. Here, workers turned the eggs just as the mother bird would have done. Turning the egg keeps the developing chick from sticking to the shell. After they hatched, the newborn chicks were fed by hand.

ornithology:
The study of birds.

When they were four weeks old, they were placed in a box on top of a tower 22 meters (about 75 feet) high. Remember, the peregrine falcon's home in the wild is usually on top of a high cliff. It was important that the birds learned to live in an environment as much like their natural environment as possible. Something else was important, too. The young birds could not have contact with humans. This would be dangerous for the birds' survival in the wild. Can you guess why? Birds that live with people are used to being fed by people. These birds may have a hard time learning how to get the food they need to survive in the wild.

At six weeks, the young falcons began to fly. Soon, the birds were ready to learn to hunt. They learned through their mistakes. There were no parents to teach them. When the ornithologists thought the young falcons were ready to begin hunting, they stopped feeding them. Once the birds began hunting for food for themselves, they were ready to be released.

Will the young falcons survive in the wild? Will they live to raise their own healthy chicks? Only time and more research will give us the answer.

✔ Check yourself

1. What kept the peregrine falcon chicks from hatching?
2. What happens to chemicals, such as DDT, as they pass through a food chain?
3. How did DDT hurt the falcons?
4. What is the study of birds called?
5. What does turning an egg do for a developing chick?
6. What two things must young falcons learn to survive?

Science in our world

From earliest times, people have been interested in birds. Feathers were once an important part of clothing and head dress. In poetry and art, birds are used as a symbol of strength and freedom. The bald eagle, a bird of prey, is the national bird of the United States.

Today, laws protect many kinds of birds. It is illegal to hunt some birds for food or for their feathers. The law does not allow people to keep some birds as pets. In general, the law protects birds that are in danger of dying out. Bald eagles and peregrine falcons are protected birds.

Careers

Wildlife conservationists are people who develop and put into practice programs to save those animals and plants that live in the wild. 1 They study the plants and animals in a certain area to learn how they behave, what their problems are, and what can be done to protect them.

The Great Swamp is a nearly 6000-acre wildlife refuge in New Jersey. The refuge is divided into two sections: the wilderness area, which is left untouched; and the management area. In the management area, wildlife conservationists experiment with different methods to make the area a better place for wildlife to live and breed.

Conservationists have set aside special places where scientists, visitors, and others can stand and study birds without disturbing them.

1

Wildlife conservationists need to have a college degree in a science such as biology. But most of all, they must love their work and be interested in protecting living things.

Review the main ideas

Although there are many different kinds of birds, all birds have some things in common. All birds hatch from eggs and have bodies that are covered with feathers. All birds have wings and beaks. Birds of prey are hunters. Excellent eyesight, strong legs, talons, and hooked beaks adapt birds of prey to a life of hunting. Peregrine falcons are skilled hunters.

About thirty years ago, these powerful birds began to vanish. Scientists looked for the cause by studying the food chains. They discovered that DDT was destroying the birds' ability to form eggshells. Now, many people are working to save the peregrine falcon. Birds hatched and raised by human beings are being released into the wild. Only time will tell if the peregrine falcon will be safe in the wild again.

Chapter review

Science vocabulary

Choose the items in column B that best match the terms in column A.

A

1. adapted
2. prey
3. talon
4. migration
5. primary consumers
6. secondary consumers
7. food chain
8. pesticide
9. ornithology

B

a. eat plants only
b. study of birds
c. seasonal movement
d. hunted animal
e. chemical used to kill pests
f. energy passes along it
g. suited to an environment
h. eat animals
i. curved claw

What did you find out?

Answer the following.

1. Peregrine falcons strike prey with their _____ .
2. Birds of prey have hooked beaks, strong feet, and keen _____ .
3. Animals get energy from _____ .
4. Energy is passed from one living thing to another along a _____ .
5. DDT is an example of a _____ .
6. Pesticides weaken a bird's _____ .
7. When an entire species dies off, it is said to be _____ .

Think and write

1. Find out what species of animals have become extinct in the twentieth century.

2. Select one of these species. Tell why it became extinct.

3. Select a bird of prey other than the peregrine falcon. Write a report about its life.

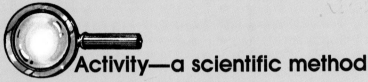

Activity—a scientific method

Purpose: to watch a bird hatch.

Materials needed

a fertile bird's egg   pencil and paper

Procedure

1. Observe the egg.
2. Note how the chick gets out of the egg.
3. Note how the chick behaves after it is out.
4. Record your observations.
5. Observe the chick up to a week old.
6. Record how it behaves at different ages.

Observations

1. How did the chick get out of the egg?
2. Why did the chick behave as it did after it hatched?
3. In what ways does a chick's behavior differ from an infant's?

Mammals

In this chapter, you will find out that

■ A warm-blooded animal is one whose body temperature stays the same.

■ Mammals are warm-blooded animals and feed milk to their young.

■ Most mammals give birth to living young.

■ Mammals live on land and in the water and are found in most areas of the world.

Before you begin

Do you know that the largest land animal is the elephant? Elephants may be as tall as 3 meters (10 feet). They may weigh more than 4,500 kilograms (about 10,000 pounds). Do you know an animal bigger than an elephant? The blue whale is the largest animal on the Earth. The blue whale may grow to be over 30 meters (about 100 feet) long. It may weigh more than 10,000 kilograms (about 22,000 pounds)! Yet, both of these animals have many common features.

The blue whale and the elephant also have features similar to bats and to a tiny mouselike animal called a shrew. All of these animals are mammals. Dogs, cats, cows, and horses are mammals, too. Anteaters, human beings, and polar bears are also mammals. In this chapter, you will learn how all mammals are alike. You will also learn about the different kinds of mammals.

What is a mammal?

Some mammals are very small. Others are very large. But no matter what their size or shape, mammals have many common features. Mammals make up a large group of animals.

All mammals are **vertebrates** (vûr′tə brāts′). Vertebrates are animals with backbones. You can feel the small bones of your backbone.

Mammals are also **warm-blooded**. This means that the temperature inside their bodies stays the same no matter what the temperature is outside their bodies. Your body temperature is always about 37°C (98.6°F).

All mammals have hair or fur. Most mammals have a layer of hair or fur that keeps them warm when the weather is cold. Some mammals, such as whales and elephants, have only a little bit of hair. But bears are covered with fur. 1 So are raccoons. 2

Most mammals give birth to living young. But did you know that some mammals lay eggs? The spiny anteater lays eggs. 3 All female mammals produce milk to feed their young.

vertebrate: Animal with a backbone.

warm-blooded: Having a body temperature that does not change with the outside temperature.

1

2

3

All mammals have lungs that they use to breathe. Even whales have lungs! They swim to the surface to breathe air.

Where mammals live

Mammals live almost everywhere on the Earth. You can find mammals on land and in water. Most mammals live in **temperate** (tĕm′pər ĭt), or mild, climates. In North America, mammals live everywhere. They even live in the cities.

temperate:
Mild climate.

Mammals live in very cold areas and in very hot areas. What animal do you picture when you think of the North Pole? You probably thought of a polar bear. Polar bears are mammals. A walrus is a mammal, too. It also lives in cold areas. 4 Some mammals, such as camels and kangaroo rats, live in the deserts. 5,6 Desert mammals can go a long time without drinking water.

4

5

6

71

Check yourself

1. Describe three characteristics of mammals.

2. Define "warm-blooded."

3. Name two mammals that live in the desert and two that live in very cold areas.

Three main groups

Mammals are divided into three main groups. Mammals in each group reproduce their young in different ways. The first group lives only in Australia and New Guinea. These mammals lay eggs. Beside the spiny anteater, the duckbill platypus (**plăt'ə pəs**) is an egg-laying mammal. 1

Have you seen a platypus at the zoo? The platypus lives near water. It is a strange animal. It has a bill that is similar to a duck's bill. The platypus has no teeth. It uses its bill to crush snails it finds on the bottom of lakes. It also eats insects. The platypus has thick fur and webbed feet.

1

Have you ever seen a kangaroo? What was in its pouch? 2 The pouched mammals are the second main group of mammals. Pouched mammals do not lay eggs. They give birth to living young.

The young are helpless at birth. They are tiny, blind, and hairless. Soon after birth, the young crawl up the female's belly to a special pouch. In the pouch, they stay warm. They feed on milk and grow. When the young mammals are large enough and can take care of themselves, they leave the pouch.

Pouched mammals live in Australia and New Guinea and also in the Americas. The opossum and the koala bear are pouched mammals. 3, 4 The largest pouched mammals are the red and gray kangaroos of Australia. They can be 1.8 meters (6 feet) tall and weigh 45 kilograms (100 pounds). Kangaroos have large strong legs and big tails. Their tails help them stand, jump, and hop.

2

3

4

The third group of mammals are those whose young are carried inside the female. They give birth to living young. These mammals do not have pouches. When the young are born, they are more developed than mammals in the other two groups. The young are fed milk by the mother. 1 This group of mammals is often called the "true" mammals. Human beings, cows, dogs, and cats are all "true" mammals.

1

Activity

Animal names

You will need reference books and other sources of information on mammals.

1. Make a chart. 2

2. List the names of these mammals in the chart: horse, cow, kangaroo, deer, goat, tiger, seal, giraffe, fox, bear, elephant, rhinoceros, pig, wolf.

3. Find the names given to the offspring for each of the mammals. Put them on the chart.

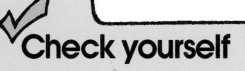

Mammals	Offspring Names
1. Horse	Foal
2. Cow	
3. Kangaroo	
4. Deer	
5. Goat	

2

Check yourself

1. List the three main groups of mammals.

2. Name two mammals that are in each group.

3. How are young mammals fed just after they are born?

4. Which group of mammals gives birth to young that are most developed?

Different kinds of "true" mammals

There are ten main groups of "true" mammals. The largest group is the **rodents** (rōd′nts). These are small land mammals. All rodents have long front teeth for tearing plants. Mice, rats, squirrels, beavers, and guinea pigs are rodents. The largest rodent is the South American capybara. 1 It is the size of a dog. It may weigh as much as 50 kilograms (110 pounds).

Hoofed mammals are another group of "true" mammals. Their feet have hoofs. 2 Many of these mammals run fast. Horses, pigs, cows, sheep, and goats are hoofed mammals. So are antelopes, rhinoceroses, hippopotamuses, and camels.

Many hoofed mammals have large stomachs. Their stomachs are divided into many parts. Their stomachs help them digest the large amount of plants they eat.

The **primates** (prī′māts′) are another group of mammals. They are the most intelligent mammals.

rodent:
Small land mammal with long front teeth, including mice, rats, and squirrels.

primate:
Mammal with grasping hands, including monkeys, apes, and humans.

1

2

Lemurs, apes, monkeys, and humans are primates. Many primates spend all of their time in trees. But larger primates, like the chimpanzee and gorilla, spend much of their time on the ground. All primates have grasping hands. They can hold things with their hands. 3

Elephants are the trunk-nosed mammals. Elephants live in forests and grasslands in Africa and Asia. Like the hoofed mammals, they also eat plants. The elephants were hunted by humans for their ivory tusks. They are now **endangered** (ĕn **dān′**jərd) animals. They are protected by law.

Have you ever been to a marine amusement park? If you have, you've seen **marine mammals**. These mammals live in the ocean. Whales, dolphins, porpoises, and seals are marine mammals. These mammals have flippers that they use for swimming. 4

The dugong and the manatee are other marine mammals. 5 Have you ever heard of them? They are large and slow-moving. They, too, have flippers that they use for swimming. Sailors used to think that these mammals were mermaids!

Marine mammals have a thick layer of oily fat called **blubber** (blŭb′ər). The blubber is under the skin. It helps keep the animals warm in cold water.

Although marine mammals breathe air, they can stay under water for a long time. Seals and walruses spend only part of their time in the water. Whales, porpoises, and dolphins never leave the water. Some scientists think that these large animals are very intelligent. Some scientists also think dolphins and porpoises can communicate with each other by sound. These sounds may be part of a language!

3

endangered: In danger of disappearing from the Earth.

marine mammal: Mammal that lives in the ocean.

blubber: 4 Layer of oily fat in marine mammals.

5

Activity

Animal communication

Most animals, including humans, communicate through body language.

1. Observe a pet dog or cat. Look at its tail, mouth, and ears.
2. Make a record of how the animal acts.
3. How do these animals show emotion?
4. Do humans communicate with body language? Give an example.

Check yourself

1. What is the largest group of "true" mammals?
2. What characteristics make horses and rhinoceroses similar?
3. Name a trunk-nosed mammal.
4. Name two marine mammals that do not leave the water.

More "true" mammals

The flying mammals, or bats, are another large group. Bats do not have feathers. They are more like flying rats than birds. Their bodies are covered with fur. 1

The insect eaters are also a group of mammals. They are usually small. Moles and shrews belong to this group. But some insect eaters, such as hedgehogs, are as big as rabbits.

The **carnivores** (kär' nə vôrz') are another group of mammals. They eat other animals. Lions, polar bears, dogs, cats, and weasels are in this group. Many carnivores hunt and kill other animals. But some carnivores, such as the hyena (hī ē' nə), eat animals that are already dead.

The rabbits and their long-eared cousins, the pikas, make up another group of mammals. 2   Did you know a jack rabbit can outrun a coyote? A jack rabbit can hop at 70 kilometers per hour (about 43 miles per hour).

The toothless mammals are a small group of strange animals. The anteaters, tree sloths, aardvarks (**ärd**'värks'), and armadillos are toothless mammals. One kind of armadillo is found in Texas. 3   It is covered with bony armor. When attacked, the armadillo plays dead. It curls into a ball.

1

2

3

✓ Check yourself

1. How are bats different from birds?
2. How fast can a jack rabbit hop?
3. Describe how an armadillo protects itself.
4. List three carnivores.

Science in our world

As herds of elephants roam from place to place in search of food, they have an effect on the environment. Because they must eat large quantities of food, they must roam over large areas.

Unfortunately, elephants must now compete with people for land. New villages and farms are pushing the elephant into smaller feeding areas. Elephants spend hours grazing on grass and eating leaves. They use their trunks to strip bark from the trees. Vegetation is destroyed.

Some countries have tried to solve this problem by establishing national parks for the elephants. Will these magnificent animals survive? Only time will tell.

Careers

Zoologists (zō ŏl′ə jĭsts) are scientists who study animals. Jane Goodall is a famous zoologist who has studied chimpanzees. 1 She has learned many things about how chimpanzees behave.

Most of what she has learned has come from observing. Scientists must be careful observers.

Goodall observed wild chimpanzees in an African forest. It was not easy. She had to spend many weeks in the forest. There, she watched the animals from a distance. Slowly the chimpanzees began to trust her. Finally they let her get close and even acted as if she were part of their group.

Goodall learned that chimpanzees use tools. They catch insects with sticks and twigs. The chimpanzees put sticks into termite holes and eat the termites that crawl onto the stick.

1

Review the main ideas

Mammals are warm-blooded vertebrates. They have hair or fur on their bodies. Female mammals produce milk for their young.

Some mammals lay eggs. The platypus that lives in Australia is an example.

Most mammals give birth to living young. The pouched mammals, such as the opossum and kangaroo, give birth to living young. The young crawl into their mother's pouch, where they feed on milk and grow. The largest group of mammals is the "true" mammals. They give birth to well-developed young that are fed milk by their mother. They do not have pouches.

Chapter review

Science vocabulary

Choose the items in column B that best match the terms in column A.

A	B
1. vertebrate	a. humans belong to this group
2. warm-blooded	b. living in water
3. mammal	c. having body temperature that does not change with outside temperature
4. carnivore	d. an animal that has hair or fur, gives birth to living young, and feeds milk to its young
5. blubber	e. an animal with a backbone
6. marine	f. oily fat of marine mammals
7. primates	g. mammal that eats mostly meat

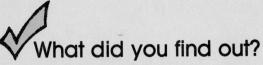 What did you find out?

Answer the following.

1. Female mammals produce _____ for their young.

2. Because mammals are _____ , they can live in cold climates.

3. The platypus lays _____ .

4. Newborn kangaroos stay in a _____ after they are born.

5. There are _____ groups of "true" mammals.

6. Whales and dolphins must breathe _____ .

7. Primates have _____ hands.

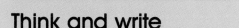

Think and write

1. Does it matter if whales become extinct? Explain your answer.
2. What are some advantages of giving birth to living young instead of laying eggs?

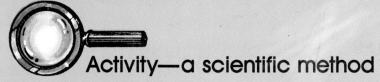

Activity—a scientific method

Purpose: to visit and observe an elephant at the zoo.

Materials needed

pencil and paper   references on elephants   a zoo (if possible)

Procedure

1. If there is a zoo near you, find out if you may visit it.
2. Go with a classmate or two. Find out where the elephants are kept.
3. Ask the zookeeper if you are seeing Asiatic or African elephants.
4. Note *what* the elephants eat and *how* they eat.
5. Notice the elephants' tusks, ears, and hair.
6. Next, read about other elephants. For example, if the elephants you saw at the zoo were Asiatic, read about African elephants.

Observations

1. Describe what and how the elephants eat.
2. Describe the elephants' tusks, ears, and hair.
3. Use your observations at the zoo and what you have read to describe the differences between the African and the Asiatic elephant.

Bones, muscles, skin

In this chapter, you will find out that

- The human skeleton is made up of bones that support and protect the body.
- Bones are attached to each other by ligaments.
- Muscles cover the skeleton and allow it to move.
- Muscles are held to bones by tendons.
- Skin is made up of layers and protects the body.

Before you begin

Val likes to dive. She practices her diving whenever she can. By practicing, Val has developed a good diving form. That means she can dive with a smooth, graceful motion.

Sometimes Val does not hit the water perfectly. When her body hits the surface, the water sprays all around her. Yet Val's body keeps its form and does not fall apart. Have you ever smacked your hand against the water? If you have, you probably felt a sting. But did your hand lose its shape or fall apart? What gives your body its shape and protects it and its organs? In this chapter, you will find out.

Holding it all together

There are over 200 bones in your body. Each bone has a special job to do. All your bones are joined to each other to form your **skeleton** (skĕl′i tən). 1   It is your skeleton that supports and protects your whole body.

skeleton:
The body framework,
consisting of 206
bones.

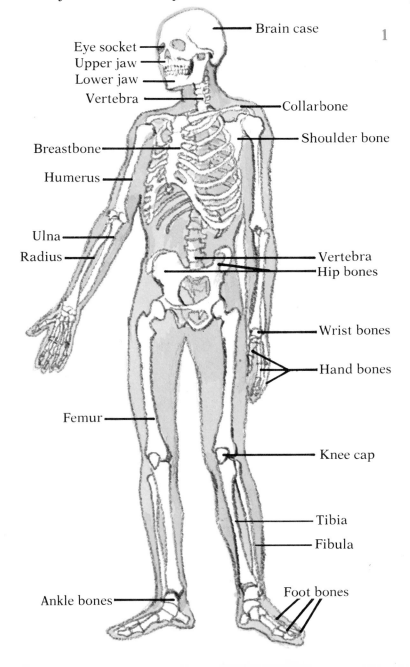

1

- Brain case
- Eye socket
- Upper jaw
- Lower jaw
- Vertebra
- Collarbone
- Shoulder bone
- Breastbone
- Humerus
- Ulna
- Radius
- Vertebra
- Hip bones
- Wrist bones
- Hand bones
- Femur
- Knee cap
- Tibia
- Fibula
- Ankle bones
- Foot bones

Activity

Name the bones

1. Find a reference book that shows the entire human skeleton.
2. Using tracing paper, draw a skeleton.
3. After you have done this, see if you can find and label these bones: femur, tibia, humerus, radius, ulna, sternum, clavicle.

Have you ever thought about how you might look without your skeleton? You might look like a bean bag, or a "pillow person." Your skeleton gives your body its shape. It helps you walk on two legs, touch your toes, and reach for a ball.

Your skeleton is important because it protects your body. Certain groups of bones form special protection for softer parts of your body. Your skull forms a tough hat to protect soft brain tissue. Your rib bones join to form your rib cage. This strong bone cage protects your heart and lungs. Your spinal column, or backbone, is made up of bones that protect your spinal cord. The backbone helps to hold you erect.

Living bones

What do you think the inside of a bone looks like? Do you think it is hard and tough? Do you think it is like a wooden log? Let us find out. Art put a chicken bone in a jar of vinegar. When he took it out in a few weeks, the chicken bone had become soft. He could bend it easily. 1 He could even tie it into a knot. Do you know why the bone became soft?

The outer layer of bone is made up of hard mineral material. 2 Calcium makes up most of this tough outer layer. The vinegar in Art's jar dissolved most of the calcium and other minerals that make bones tough. That is why Art was able to bend the bone so easily.

Your bones are made up of living material that is changing constantly. Some of the calcium from your bones is used by your body. Then, new calcium is used to form new layers of bone. Bones are only hard and strong on the outside. Inside the bone, there is a soft protein material called **bone marrow** (bōn **mǎr'ō**). Most of the body's red blood cells are made in the bone marrow.

All bones do not have the same amount of hard mineral material. Usually, the ones that have hard work to do have a thicker outer layer and less bone marrow than some other bones. For example, the

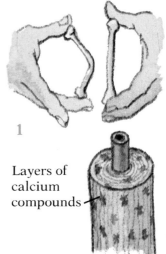

1

Layers of calcium compounds

Cross Section of Bone

2

bone marrow:
The soft protein material found inside the bone.

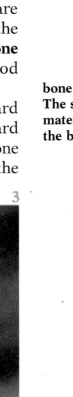
3

88

thighbone, or **femur** (**fē′**mər), is the strongest and largest bone in your body.

You have a kind of soft bone, too. Did you ever press your nose against a store window? 3 You probably looked very funny. You were not afraid you would spoil the shape of your nose. You knew you would look just like yourself again. Your nose can bounce back because its tip is made of a firm, elastic material called **cartilage** (**kär′**tl ĭj). Cartilage can bend and twist easily. It also gives your ears their shape. When your bones grow longer, you grow taller. As a new bone forms, cartilage is broken down and leaves the area. The new bone takes the place of the cartilage.

femur:
The largest bone in the human body; the thighbone.

cartilage:
The firm but elastic material that makes up the nose and other parts of the body.

✔ Check yourself

1. How many bones are in your body?
2. What does your skull protect?
3. What mineral material makes up most of a bone's outer layer?
4. What is the soft material inside the bone?
5. Where are most of the body's red blood cells formed?
6. Name the largest bone in your body.
7. Describe the cartilage at the tip of your nose.

Holding your bones together

If most of your bones are strong and hard, how can you bend your arm? How can Mary bend her body like this? 1   You are able to bend because of your **joints**. The place where two bones meet is called a joint. There are three different kinds of joints in your body. The first kind of joint does not move at all. You may be surprised to know that your skull is made up of many different bones. Your skull bones have joints that do not move.

Gliding joints

Another kind of joint moves just a little. Do you know what part of your body has joints like these? Run your fingers down the middle of a friend's back. Can you feel the small bones that make up the spine? These bones are called **vertebrae** (vur′tə brē). 2   Can you bend to touch your toes? When you do, each joint between your vertebrae moves just a little.

There are other joints that can move much more than your vertebrae can. The two types of joints that are most movable are the ball and socket joint and the hinge joint. Put your left hand on your right shoulder. Now, raise your right arm and swing it in large circles. Your shoulder is a ball and socket joint. This joint is important for athletes who play baseball and basketball. 3 Your knee and your elbow are hinge joints. Hinge joints move only in one direction, not in a circle. What hinge joints are the runners using? 4

Tying your joints together

Have you ever seen a dancing skeleton at Halloween? Why did its arms or legs not fall off? Its joints were probably held together with string. In your body, your movable joints are also held together. They are held together by strong, elastic fibers called **ligaments** (lĭg′ə mənts). Ligaments can stretch. They allow the bones to move freely. Sometimes, when we play too hard or fall, we stretch our ligaments too far. Did you ever sprain your ankle? If you did, you may have pulled your ankle ligaments until they snapped.

3

ligament:
**The elastic material
that holds together
the movable joints of
the body.**

4

✔ Check yourself

1. What is the place where two bones meet called?
2. Which small bones make up the spine?
3. What are the most movable joints in your body called?
4. What fibers hold together your movable joints?

Making your move

You have learned that the body has movable joints. But bones cannot move by themselves. They need help. How do you stand up or sit down? How do you comb your hair, or ride a bicycle? It is the muscles in your body that move your skeleton and other parts of your body. The muscles and bones work together so you can move. Muscles cover the bones. They fill out your body and give it its shape. Have you ever seen a weight-lifter? What do you see bulging under the weight-lifter's skin? These are the weight-lifter's muscles. They have been built up by regular exercise.

When you move, your muscles bunch up and get smaller. This is called contraction. Then, the muscles relax, or go back to their normal size. All the muscles in your body move this way. But there are different kinds of muscles. Some muscles are under your control. Others are not. The muscles that are not under your control are called involuntary muscles. When you eat, the muscles that push your food down to your stomach are involuntary muscles.

The kinds of muscles that are under your control are called voluntary muscles. Raise your hand. Wave it in the air. You just used your voluntary muscles. These muscles are attached to your bones by narrow, strong cords called **tendons** (tĕn′dənz). Open your hand and spread your fingers apart. Now, if you look at the back of your hand, you will see some of your tendons. They go from your knuckle to your wrist. Touch them to see what they feel like. You will find that they are hard and strong.

When your muscles contract, they pull the bone. But muscles cannot push the bones. They work in only one direction. Another muscle must contract to pull the bone in the opposite direction. When you bend your elbow, one set of muscles pulls your arm towards you. When you straighten your elbow, another set of muscles pulls your arm away from you. 1

Some voluntary muscles do not pull bones. The muscles that move your lips are muscles like that. When you move the muscles in your face, how do you look? You can smile, frown, or look surprised. Maybe you would like to make funny faces. You are able to do all these things because of your face muscles.

Your muscles and bones work together to support you and to help you move. But we are not the only ones with these systems. Cows, sheep, chickens, and all other animals with skeletons have muscles attached in the same way. Let us take a look at their bones. The next time you have chicken for dinner, get the drumstick for examination.

tendon:
Narrow, strong cord that attaches muscle to bone.

1

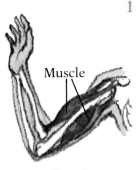

Muscle

Muscles That Move the Arm

Activity

Look at the drumstick

1. Before you bring it to class, rinse the drumstick in cold water and put it in plastic wrap. You should be able to recognize certain parts of the drumstick.

2. The meat of the drumstick that you ate for dinner is actually muscle. Do you find any shiny bands or stringlike parts on what is left? What do you think they are?

3. **Caution:** Be careful using a knife. Cut away any meat that is left around the bone. Describe this bone. Is it thick or thin? Why is this bone important in a chicken's body?

4. At the end of the bone is that tough part that you cannot eat. Peel that away with a knife. Examine it. What is it made of? You are exposing an important joint in the chicken's body. Did you see any ligaments?

5. After you clean the bone thoroughly, you may wish to place it in a jar of vinegar for a couple of weeks. What will happen to it?

If you examined your drumstick carefully, you probably noticed the tiny bands, or tendons, that attach the muscle to the bone. The drumstick must be fairly thick to support the chicken. The end of the bone is covered with cartilage. It cushions the area of the joint that allows the chicken to move.

✔ Check yourself

1. What moves your skeleton and other parts of your body?

2. What are voluntary muscles?

3. What are involuntary muscles?

4. Name the cords that attach muscle to bone.

5. Describe how your muscles work when you bend and straighten your elbow.

Wrapping you up

Have you ever received a present wrapped in beautiful paper? Sometimes, the package is too pretty to open right away. Your skin is like your wrapping paper. It hides all the wonderful things that make up your body. It is smooth and nice-looking. But your skin is more than a pretty wrapping.

Your skin protects your body in many different ways. It is sensitive to heat, cold, pain, and pressure. If this were not true, you might keep your hand near a flame and be burned. Your skin senses the danger, and you pull away. Your skin also protects your body from losing blood and other important fluids. It protects you from injuries and hard bumps. Your skin keeps dirt, fluids, or other material from getting into your body, too.

Skin grows and changes everyday. It is made up of layers. The top layer, the layer that we see, is the dead skin. Here is the way your skin grows. The top skin dries up and is rubbed off. New skin that has grown underneath replaces the old. When this becomes old, it dries up and is rubbed off, too. The deeper layers of the skin contain tiny blood vessels. Have you ever had a paper cut or a scrape? This kind of injury usually does not bleed. That is because the deeper layers of skin were not hurt.

Your skin contains **sweat glands**. 1 They are small tubes that produce a salty fluid. This fluid is called sweat. When you are warm, the sweat glands produce extra amounts of fluid. The fluid runs onto your skin through tiny openings, called **pores**. When the moisture on your skin dries, or evaporates, it helps to make you cool.

The deeper layer of skin also contains the roots of your hair. 1 Only the root of your hair is living. Have you ever pulled a hair out? How does it feel? The pain tells you that part of your hair is alive. Your hair grows from the root and pushes out above the skin. The hair itself is not living. That is why a haircut does not hurt you. The skin also contains oil. This oil keeps your hair and skin from getting too dry.

sweat gland:
Small tube found in the skin that produces salty fluid.

pore:
Tiny skin opening that releases sweat onto the skin's surface.

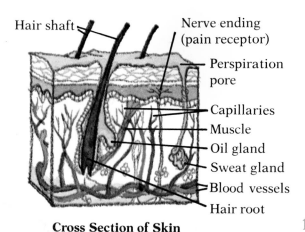

Hair shaft
Nerve ending (pain receptor)
Perspiration pore
Capillaries
Muscle
Oil gland
Sweat gland
Blood vessels
Hair root

Cross Section of Skin   1

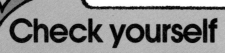

Check yourself

1. How does skin protect you?
2. Which skin layer is made of dead skin?
3. Name a small tube in the skin that produces salty fluid.
4. What are the tiny openings in the skin called?
5. What part of your hair is living?

Science in our world

Broken bones are not any fun. Have you ever seen anyone wearing a cast? Sometimes the broken bone is put in a plaster cast to keep it from moving around. The person cannot use that part of the body for a long time. It can be very uncomfortable to wear a cast. It takes about four to six weeks for most bones to heal.

Scientists are working on a new method to help bones heal. They have found that electricity will help bones to grow. Electric current can even help heal bones that would not heal in any other way. Scientists have also found ways of replacing parts of bones with metal or plastic. People who could never be helped before are now able to run and play again.

Careers

Those who help people suffering from joint and bone diseases are called **physical therapists** (fĭz'ĭ kəl **thĕr'ə** pĭsts). 1 Their patients include accident victims, handicapped children, and disabled older persons. Physical therapists perform tests that measure the muscle strength and other physical abilities of the patients. Then, they develop a program of treatment in cooperation with doctors and nurses. Usually, physical therapists work in hospitals. But they may work in nursing homes and schools for handicapped children.

Physical therapists must have a state license to practice. A college degree or a certificate of training is also required.

1

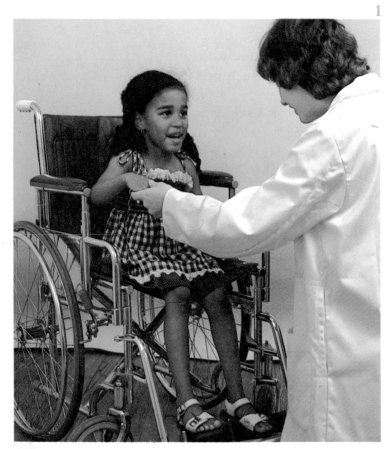

Review the main ideas

There are over 200 bones in your body. They are joined together to form your skeleton. Your skeleton supports your body. It also helps to protect the softer parts inside your body. The skull protects your brain. The rib cage protects your heart and lungs. The places where two bones meet are called joints. Bones are held together by ligaments.

Muscles cover the skeleton and allow it to move. Muscles are held to bones by tendons. There are two different kinds of muscles. The first kind of muscles is under your control. They are called voluntary muscles. These include the muscles in your arms and legs. Your face muscles are also voluntary muscles. The second kind of muscles is not under your control. They are called involuntary muscles. Your heart is an involuntary muscle.

Your skin protects your body. It is made up of layers. The top layer is the one we see. It is made up of older skin. When this skin dries up, it peels away. The deeper layers of skin contain tiny blood vessels, sweat glands, hairs, and oil glands.

Chapter review

Science vocabulary

Choose the items in column B that best match the terms in column A.

A	B
1. skeleton	a. a firm but elastic material
2. bone marrow	b. consists of over 200 bones
3. cartilage	c. holds bones together
4. vertebrae	d. soft inside of bone
5. tendons	e. attaches muscle to bone
6. ligaments	f. small bones of the spine

What did you find out?

Answer the following.

1. Your body gets its support from your _____ .
2. The outside part of bone is made of _____ .
3. Muscles are attached to bone by _____ .
4. An example of a movable joint is your _____ .
5. The bones in your body number _____ .
6. Muscles that you can control are _____ muscles.

Think and write

1. Explain why it is necessary to have a skeleton.
2. Why do muscles often work in pairs?
3. Describe one of your activities that involves involuntary muscles.

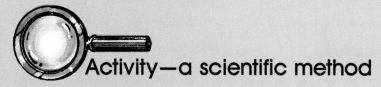

Activity—a scientific method

Purpose: to count the joints in your body.

Materials needed

pencil and paper   a partner

Procedure

1. Work with a partner. This will let you compare observations.
2. Try to count the number of joints in your body. You will only be able to count those that move.
3. Start with your hands. Count the number of joints in your fingers. Record the number.
4. Count the number of joints in your arms and legs.
5. Count the number of joints on your torso. (Your torso is between your thighs and your neck.)
6. Count the joints in your partner's back. Let your partner count the number of joints in your back. You should have the same number.
7. Count any joints you can find in your head.
8. Count the joints in your legs and feet.
9. Record all the joints you find. Also record where you find them.

Observations

1. How many joints did you find altogether?
2. How many joints did you find in your head?
3. Why could you not find more joints in your head?
4. Where did you find the most joints? Why are they there?

7

The eyes and vision

In this chapter, you will find out that

- We see things because light bounces off them into our eyes.
- The pupils of the eyes change size to let in more or less light.
- Certain cells on the retina detect colors, and others detect shapes.
- Moving pictures are optical illusions.
- Eyeglasses correct some vision problems.

Before you begin

Have you ever painted a still life? Ali is doing that in the picture. Does it look easy? A lot happens between the time that Ali looks at the flowers and the time she paints them. Many of those things are happening in her eyes. Still others are happening in her brain.

Light bounces off the flowers and into Ali's eyes. Certain parts of her eyes see the shapes of the flowers. Still other parts see the color. Ali does not actually see the flowers in her eyes. Do you know where seeing really takes place? Did you know that if Ali were to draw the flowers as they appear to her eyes she would draw them upside down? Do you know how many parts of your eye must work together for you to see?

Letting in light

Close your eyes for a moment. Think how different your world would be without light. For you to see, light from the thing you are looking at has to enter your eyes. Light first enters the eye through the clear, front part of the eyeball. This protective covering is called the **cornea** (kôr′nē ə). Then, the light goes through an opening called the **pupil** (py$\overline{oo}$′pəl). The pupil is covered by the cornea and is the dark-looking part at the center of the eye. Around the pupil is the **iris** (ī′rĭs).

The iris gives your eye its color. It also controls the amount of light that enters through the pupil. As the brightness of the light changes, the size of the iris changes. And when the iris changes size, so does the pupil. When the pupil is larger, more light can get into the eye. 1   When the pupil is smaller, less light can enter. 2

cornea:
The clear, protective covering in the front of the eyeball where light first enters the eye.

pupil:
The dark-looking part in the center of the eye through which light passes.

iris:
The colored part of the eye that surrounds the pupil and controls the amount of light entering the pupil.

1

2

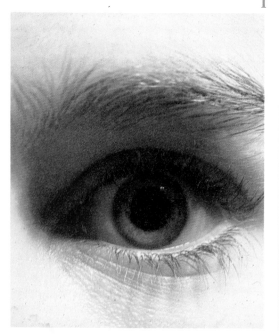

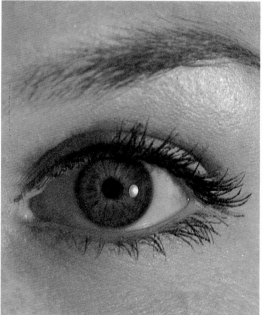

 **Activity**

Watch your pupils change

1. You need a partner, and a watch or a clock.
2. Hold your hands over your eyes and keep them shut. Have your partner time you for 1 minute.
3. Take away your hands and let your partner watch your pupils as they change.
4. What does your partner observe? Trade places. This time you watch your partner. What do you observe?

It takes time for your eyes to adjust to changes in light. In the activity, the iris expanded, causing the pupil to get smaller as light began to enter the eye. The opposite happened while your hands covered your eyes. In the darkness, the iris contracted and caused the pupil to get larger.

✓ Check yourself

1. What must enter your eyes for you to see?
2. Why is the cornea important?
3. Name the dark-looking part in the center of the eye.

Images in the eyes

Light that goes into the pupil of the eye passes through the **lens** (lĕnz). The lenses in your eyes are curved, so they are thicker in the middle than at the edges. The lens bends light, as shown in the diagram. 1

lens:
Something that bends light.

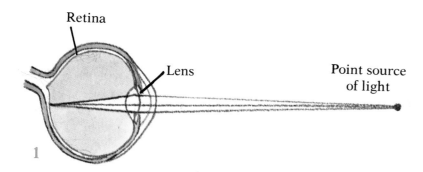

Retina

Lens

Point source of light

1

The bent rays of light come together and form an image, or picture, on the **retina** (rĕt′n ə) at the back of the eye. We say the lens **focuses** (fō′kə sĭz) the light on the retina. When something is in focus, we know the image is clear and sharp.

Look at the diagram, which shows the image of a tree on the retina. 2 Why is the image upside down? Follow the ray of light that goes from the top of the tree. When it strikes the lens, it bends and changes direction. As a result, the image of the top of the tree forms at the bottom of the retina. Now, follow the ray of light from the bottom of the tree and see how it bends and changes direction, too. Light from the bottom of the tree forms an image at the top of the retina. If you could look inside the eye, you would see a tiny, upside-down tree. Then, why does the world not look upside down?

retina:
The back part of the eye on which an image forms.

focus:
To make clear and sharp, as in a focused picture.

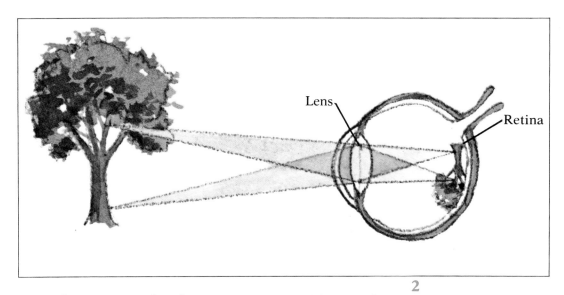

Look again at the diagram on page 106. At the
back of the eye, there is a nerve that goes to your
brain. The cells of the retina change light into
signals that go through the nerve to the brain. The
brain turns the image over and makes it right-side
up again. It is really the brain that sees things.

Focusing near and far

The lenses in your eyes change shape. Muscles
around each lens pull tight to make the lens thin.
The muscles relax to make the lens thick. 3 The
lens becomes thinner to focus light coming from
things that are far away. It becomes thicker to
focus light coming from things that are close by.

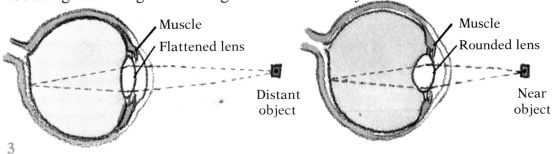

Eye Muscles Change the Shape of the Lens

Check yourself

1. What part of your eye focuses light to form a picture?
2. Where is the picture formed?
3. Explain why it is the brain that really sees things.
4. How does the lens focus light?
5. Describe the shape of the lens when you look at something that is far away.

Color and the eye

Most people have three kinds of color-detecting cells. Each cell detects, or discovers, one of the three basic colors—red, green, and blue. These color cells are called **cones,** because they are cone-shaped. Some people do not have all three types of cones. These people are colorblind. For example, a colorblind person might not see the number in this picture. 1 The person cannot see those colors because the detecting cells are missing.

Other cells in the retina detect very dim light. These are called **rods,** because they are rod-shaped. In dim light, only the rod cells work. Rods permit you to see something's shape. Have you ever looked at an object in a dark room? Often, you can see its shape but not its color.

cone:
Cell in the retina that detects color.

rod:
Cell in the retina that detects dim light.

108

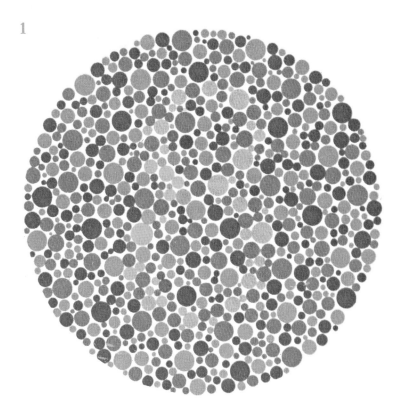

Seeing what is not there

Watch a person walk slowly across the room. While you are watching, blink rapidly. What do you see?

The eye takes a very short time to see an image. During this time, a blink makes no difference in what you see. You may even see something for a short time after it has gone. Because of this, you may sometimes see things that are really not there. Scientists call this an **optical illusion.**

There are different kinds of optical illusions. In this kind, the retina continues to send messages of an image to your brain for a short time after the object has gone. Find out about another kind of optical illusion by doing the activity.

optical illusion;
Something that appears to be what it really is not.

Activity

Optical illusion

1. You need a darkened room, a flashlight, and a partner.
2. Turn on the flashlight and hand it to your partner. Stand about 2 meters (about 6½ feet) away.
3. Ask your partner to swing the light around in a circular motion. What do you observe?

When the flashlight is swung around, you see an image of a complete circle of light. 1 But you know that what was really there was a single spot of light moving around. The image of each spot fades slowly enough so that all the earlier spots of light seem to still be there.

1

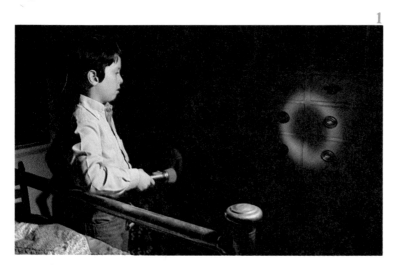

Moving pictures are an optical illusion. Each frame, or picture, in a motion-picture film differs by a small amount of motion. 2 Motion takes place between the times the pictures are taken.

2

A motion-picture projector shows one picture at a time. The shutter in the projector cuts off the light as the film moves. The projector moves the film ahead, frame by frame. You see motion because your brain fills in the missing movement.

Check yourself

1. What do cones do to help you see?
2. What are the three kinds of cones?
3. What makes some people colorblind?
4. What do rods do?
5. What is an optical illusion?

Science in our world

Many people have different kinds of problems with their eyes. They may wear eyeglasses to help correct these problems. We do not know when eyeglasses were invented, but there are reports of their being used 6000 years ago. Benjamin Franklin invented the bifocal (bī **fō′**kəl) lens about 1760. The bifocal lens is made of two lenses joined together. The upper lens is used to see things far away. The lower lens is used to see things up close.

Some people are nearsighted. Nearsighted people can see clearly only those things that are near their eyes. This is because the image focuses in front of the retina, making objects far away look blurry. 1 Other people are farsighted. Farsighted people see best the things that are far away. This is because the image focuses behind the retina. 2 Both nearsightedness and farsightedness can be corrected by eyeglasses.

Lenses that rest on the eyeball are called contact lenses. Today, many people wear contact lenses. Athletes wear them. So do workers who cannot wear glasses because the glasses might break on the job. Contact lenses are made of plastic.

1

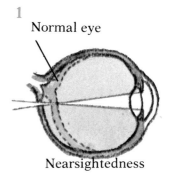

Normal eye

Nearsightedness

2

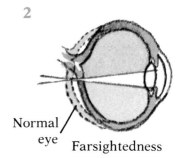

Normal eye

Farsightedness

Careers

Sight is one of our most important senses. But many people have difficulty seeing clearly. They must wear eyeglasses or contact lenses to help correct their vision problems. Helping people to see better is an interesting and important job.

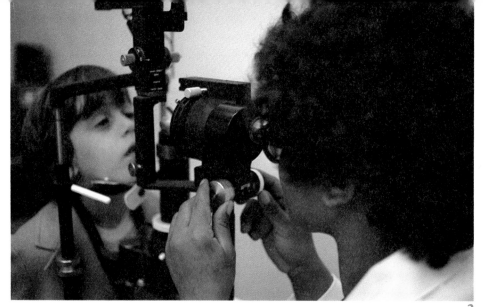

Optometrists (ŏp′tŏm′ĭ trĭsts) examine eyes for vision problems. **3** They test the person's ability to focus and to judge distance and color. They prescribe lenses to correct sight problems.

Ophthalmologists (ŏf′thăl mŏl′ə jĭsts) are doctors who specialize in eye care. They perform eye surgery and prescribe lenses.

Opticians (ŏp tĭsh′ənz) help people choose eyeglass frames and contact lenses. They also grind lenses. When they do this, they follow prescriptions written by optometrists and ophthalmologists.

Review the main ideas

When we see an object, light first bounces from the object into the eyes. The pupils change size according to the brightness of the light. Going through the lens, the light focuses on the retina, where an upside-down image forms. Signals go through a nerve to the brain, where the image is seen right side up. Cones and rods on the retina detect colors and dim light. Moving pictures are a kind of optical illusion. People use eyeglasses and contact lenses to correct vision problems.

Chapter review

Science vocabulary

Choose the items in column B that best match the terms in column A.

A	B
1. rod	a. clear, front part of the eyeball
2. cone	b. colored circle with hole in it
3. pupil	c. gets smaller in bright light
4. iris	d. where an image is formed inside the eye
5. retina	e. detects color
6. cornea	f. detects dim light

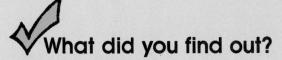

What did you find out?

Answer the following.

1. The lens of the eye _____ light.

2. In bifocal glasses, two _____ are joined together.

3. When the eye moves into dim light, the _____ gets larger.

4. People who do not have all three kinds of cones are _____ .

5. You see an optical _____ when you swing a light in a circle.

Think and write

1. How are eyes like a movie camera? How are they different?

2. The retina has just three kinds of color-detecting cells. How can you see many more colors than just three?

3. Can you always believe your eyes? Explain why or why not.

114

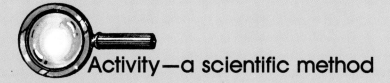

Activity—a scientific method

Purpose: to judge distances.

Materials needed

2 marbles   1 dowel   ruler
2 cardboard strips   a little clay

Procedure

1. Sit across a table from your partner.

2. Attach a marble to the center of each cardboard strip. Use clay to make the marbles stick.

3. Place the dowel on the table. One end should point toward you. The other end should point toward your partner.

4. Sight along the stick with one eye covered. Your partner should place one marble on either side of the stick.

5. Tell your partner how to move the marbles at his or her end of the table. Try to have your partner line up the two marbles. When you think they are directly across from each other, say "Stop."

6. Your partner should measure and record how far behind one marble the other is. Then, you may try again.

7. Try twice with each eye. Then, try twice using both eyes together.

8. Change places with your partner, and repeat the experiment. This time, you move the marbles for your partner.

Observations

1. Which eye worked better at judging distances, your left eye or your right eye? Which way worked the best?

2. List three occupations in which people need to judge distance.

Keeping fit

In this chapter, you will find out that

- Regular exercise helps your body to be strong and healthy.
- Your body needs a certain amount of sleep each night.
- You can keep your weight just right for you by eating the number of calories your body needs to work and play.
- The way you feel inside often has something to do with the way you act outside.

Before you begin

Look at the skater. Do you think she is having a good time? She is a speed skater. That is, she races against the clock. She also races against other skaters. To do this, she must be strong and healthy.

Exercise is an important part of keeping healthy. But it is only one of the things you should do to be physically and mentally strong. What other things can you do? Do you pay attention to what you eat? Would it be healthy to eat only your favorite food? Why does your body need food?

Rest is another important part of keeping healthy. Why do you think this is so?

117

How to be your own coach

Speed skaters have coaches who help them exercise every day. But you can be your own coach. You can run, jump, play your favorite sport, or do **calisthenics** (kăl'is **thĕn'** iks), to keep in shape. Exercising regularly will help keep you strong and healthy. Whenever you play your favorite sport, you are having fun and also are helping your body to build strong muscles.

If you cannot play your favorite sport every day, you can make up your own exercises. Can you touch your toes without bending your knees? How long can you jog without getting tired? The more you exercise, the stronger your muscles become. You will also be able to stretch more easily. It is often more fun to exercise with someone else. Try doing your calisthenics with some friends. 1

Are you getting the right amount of exercise? How is your **endurance** (ĕn **door'**əns), or staying power? You can test your endurance by checking to see how quickly your **pulse** (pŭls) slows down after

calisthenics:
A series of regular exercises.

endurance:
The ability to work for a long time without tiring.

pulse:
The beating of certain blood vessels under the skin.

you exercise. Your pulse rate is similar to your heartbeat rate. Your pulse is caused by the beating of certain blood vessels under the skin. Exercise causes your blood to flow more quickly than when you are at rest. This increases your pulse rate and heartbeat rate. The more endurance you have, the more quickly your heartbeat rate returns to normal after you stop exercising. Endurance is especially important to speed skaters, runners, and other athletes. **2**

2

✔ Check yourself

1. Define "calisthenics."
2. How do calisthenics help your body?
3. Define "pulse."
4. What is the importance of endurance?
5. What does exercise do to your blood flow?

Time to sleep

Do you know that an animal called a sloth sleeps for 20 hours a day? Giraffes and elephants sleep only four hours each day. The porpoise does not sleep at all!

But human beings need rest, or they will not have energy to work or to play. The skater whizzing down the track must be in good condition. She must be sure to get plenty of sleep each night.

Newborn babies sleep almost all the time. As we grow older, we need about 8 hours of sleep each night. Do you know why? How long do you sleep at night?

When you are tired, your body lets you know in different ways. Do you ever yawn? 1 Yawning is a signal that your body needs more energy. A quick temper can be a sign that you are tired, too. Are you ever impatient, or quick to get angry at a friend? Have you ever daydreamed in class, instead of listening to the teacher? When you do not get enough sleep, it is hard to pay attention. Your body cannot function properly without sleep. The right amount of rest will keep you feeling full of energy all day.

You just learned that most people need about 8 hours of sleep per day. (A day contains 24 hours.) Do you get enough sleep? What do you do with the other 16 hours?

Activity

How do you spend your time?

1. Keep a record of how you spend your time. Do it over a full week. Record the number of hours you spend eating, reading, and watching television. How many hours a day do you exercise? Exercising includes sports, bike riding, and other physical activities.

2. Keep your records in two categories: school days and nonschool days. Average the number of hours per day in each activity. For school days, divide the total hours by 5. (You go to school 5 days per week.) For nonschool days, divide each total by 2.

3. Make two bar graphs: one for school days, and one for nonschool days. Each bar should name a type of activity. Two boxes high should equal 1 hour. Do not forget to count sleep.

4. Which activity do you do most? Which do you do least?

5. Make sure that your graph totals 24 hours.

Your longest bar was probably sleep. Was it long enough? Do you get enough exercise? Do you watch too much television? How does your school-day graph compare to your nonschool-day graph?

✔ Check yourself

1. Do all animals sleep the same number of hours?

2. Name an animal that does not sleep.

3. Why do you need rest?

4. How many hours of sleep each day do most people need?

5. What signs indicate that your body needs more sleep?

Standing tall

Posture (pŏs′chər), or the way you stand, is very important for good health. Standing correctly helps your **spine.** The column of bones in your back must grow straight as you grow taller. Bad posture can cause backache or neck pain, too.

It is not difficult to have good posture. Good posture means standing tall with your shoulders

posture:
The position in which the body stands.

spine:
The column of bones in the back, also called the backbone.

relaxed. It does not mean standing stiff as a wooden soldier. When someone has good posture, you can tell. They just look good. Which picture shows good posture? 1, 2

Weighing in

How much you weigh is influenced by your bone structure. But your weight also depends on how much you eat and how active you are. You have to find the weight that is right for you.

Do you want to gain weight, lose weight, or stay the way you are? **Calories** (kăl'ər ēz) tell us the amount of energy in the food we eat. Since you are still growing, it is important that you eat a certain amount of calories each day. If you do not eat enough calories, you may lose weight. If you eat too many calories, you may gain more weight than you should. The number of calories for the average ten-year-old is 2400. This chart shows how many calories are in some of your favorite foods. 3

calories:
The amount of energy contained in food and released in the body.

CALORIE CHART   3

Name of food	Number of calories per serving
apple	80
banana	100
bread	70/slice
cookie	50/cookie
hamburger	185
hot dog	170
ketchup	15/tbs
milk	150/glass
mustard	5/tsp
peanut butter	95/tbs
pizza	145/slice
roll	120

Do you like hamburgers? How many calories are in one broiled hamburger? If you like to eat your hamburger on a roll with ketchup, you have to add even more calories to your total. Start to look at the packages that food comes in. 1 You will find that they usually tell you how many calories there are in one serving of the food. Try to count the calories that you eat in one day. List everything you eat at meals, and do not forget your snacks.

Activity

Count the calories

Look at what three students had for lunch.

Jose
2 slices of pizza
1 glass of milk
1 medium apple

Jennifer
1 peanut butter sandwich (2 tablespoons of peanut butter; 2 slices of white bread)
2 glasses of milk
1 medium banana

Mike
1 hamburger on two slices of bread
1 glass of milk
2 chocolate chip cookies

1. Use the calorie chart to count the number of calories each had for lunch.

2. Who had the most calories?

Those calories sure add up! If you said Jennifer had the most calories for lunch, you were right.

Feelings are fine!

Can you tell how the boy in this picture is feeling? 2 It is difficult to hide your **emotions** (ĭ **mō′**shənz), or strong feelings, from others. When people see you smiling, they think you feel good. Usually, your friends and family can tell if you are feeling angry or hurt.

Everyone has emotions. Babies cannot speak, but you know when they are happy or sad. They do not try to hide their feelings. But sometimes, older people do. We may hide our feelings because we think others will not understand. But talking about how you feel can help solve problems. Sharing your feelings makes you feel better. When you feel good inside, you are able to act your best outside.

emotion:
A strong feeling, such as anger, hate, or love.

2

✓ **Check yourself**

1. Why is good posture important?
2. What can bad posture cause?
3. What tells us the amount of energy in food?
4. What can happen if you eat too few or too many calories?
5. What are emotions?
6. Why is it important to share feelings?

Science in our world

More and more people are realizing the importance of keeping physically fit. How many people in your community work out at the school gymnasium? Many adults are jogging and swimming. Others play tennis or bowl to keep in shape. Many people are doing calisthenics several times a week.

Have you ever been in a walk-a-thon? 1 Local charities use people's interest in physical fitness as a good way to raise money. They may use walk-a-thons, jogging, or bike riding to get many people involved. Businesses are finding that physical fitness pays off, too. Workers who exercise during the day do a better job. Many companies are starting to provide time and space for their people to keep physically fit.

1

Careers

The field of physical fitness covers many kinds of jobs. Those who teach physical exercises and acrobatics are called **gymnastic instructors.** 2

Their work combines fitness and fun. It consists of teaching people how to stretch and bend the body so that it is free to move. It includes teaching acrobatic movements, such as tumbling, back bends, high kicks, and somersaults. Movements on the trapeze bar are also taught.

Gymnastic instructors work in gymnasiums, health clubs, and even schools. They may be trained in physical education, in ballet, or in modern dance. Depending on where they teach, they may need a college degree.

126

2

Review the main ideas

Regular exercise is important for everyone. It helps to build strong muscles and endurance. You should give your body a workout everyday. Both your heartbeat rate and your pulse rate go up when you exercise. How quickly they slow down after you exercise is one way of telling whether or not you are exercising enough.

Getting the right amount of sleep is also important for good health. Newborn babies sleep almost all the time. As we grow older, we need about 8 hours of sleep every day.

Proper posture and weight are also part of being fit. Calories measure the amount of energy that is in the food we eat. The average ten-year-old needs about 2400 calories each day.

Feelings are part of any physical fitness program. Everyone has emotions. When you understand your emotions, you feel better inside. When you feel better inside, you are able to do your best outside.

Chapter review

Science vocabulary

Choose the items in column B that best match the terms in column A.

A   **B**

1. calisthenics   a. the backbone
2. pulse   b. amount of food energy
3. posture   c. series of regular exercises
4. spine   d. feelings
5. calories   e. position in which the body is held
6. emotions   f. the regular beating of certain blood vessels under the skin.

✓ What did you find out?

Answer the following.

1. A series of exercises that athletes often do are called _____ .
2. Regular exercise can build up your _____ , or staying power.
3. Most people need about _____ hours of sleep each night.
4. Young people need about 2400 _____ each day.
5. Everyone has the same bone structure. (True/False)
6. Posture has no effect on your health. (True/False)

Think and write

1. Name three forms of healthy exercise.
2. What three things can happen if you do not get enough sleep?

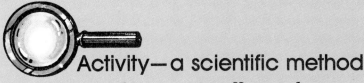

Activity—a scientific method

Purpose: to see how exercise affects pulse rate.

Materials needed

watch or clock with second hand   a partner

Procedure

1. Relax. Sit still for at least 3 minutes.

2. Open the palm of your left hand. Face that palm toward the floor. Turn your right hand over. Its palm should face the ceiling.

3. Bring your left hand to your right wrist. Place the middle and forefinger beneath your wrist.

4. Your pulse is near the tendons you feel sticking out. Press very gently. You should feel it. Have your partner time you. Count the pulses for 60 seconds. That is your pulse rate. Record it.

5. Run in place for 3 minutes.

6. Immediately take your pulse again. Record it.

7. Sit for 2 minutes. Take your pulse again. Record it. Repeat this step two more times.

8. Change places with your partner. Help your partner to perform steps 1 through 8.

Observations

1. How does exercise affect your pulse rate?

2. How long did it take your pulse rate to return to the "rest-rate"?

3. How does what you found compare to what your partner found?

4. What does your pulse rate measure?

What's in your food?

In this chapter, you will find out that

- Most of the foods you buy have substances added to them.
- Things are added to foods to make them look or taste better, to keep them fresh longer, or to make them more healthful.
- There are laws to help make sure that the foods you buy are safe for you to eat.

Before you begin

Rice blown into light puffs, apples and butter dyed to look brighter, meat that stays fresh for weeks—foods today are very different from those people used to eat. Do you know why peanut butter is so smooth and creamy? How can plain water be made to fizz? What can be done to white bread to make it as healthful as brown bread?

Many strange things are being added to packaged foods. Some of these things could be bad for your health. So people have asked to have laws passed. These laws require food companies to test the foods they make. The companies must be sure that the foods sold to us are safe to eat.

In this chapter, you will learn something about food additives. What you learn may help you make better decisions about what you eat.

Then and now

In the old country store, crackers were kept in a big barrel. Other barrels held apples, potatoes, and pickles floating in salt water. Smoked hams, sides of bacon, and big round cheeses hung from the ceiling. Flour and sugar were in cloth sacks. 1

1

Most people lived on farms where they grew their own fruits, vegetables, meat, eggs, and milk. After people learned how to can food, they could store summer foods and eat them through the winter.

When women went to work outside the home, they wanted foods that needed little time to prepare. Foods packaged in boxes, cans, and bottles were very different from foods for sale in the old country store. These prepared foods contained substances that made it possible to keep them stored for a long time. 2

2

Preserving foods

Bottled drinks that fizz, such as soda, are very popular. Here is how they are made. Take plain water. Then, add sugar, flavoring, and coloring. Finish by putting in gas bubbles (carbon dioxide gas) and something to keep the soda fresh for a long time. Everything you added to the water is called an **additive** (ăd′ĭ tĭv). Most of the food we eat has many additives. They are added to the food to make it look better, taste better, or last longer.

The first **preservative** was probably discovered before written history. That preservative was probably salt. People discovered that salt would preserve meats and fish. **3** They began to like the taste of salt on food. People living near the coast could get salt from sea water. Other people had to find and dig for salt in salt mines. As you can imagine, salt was very much in demand and very valuable.

additive:
Something that does not occur in a food naturally.

preservative:
An additive that keeps a food from spoiling quickly.

3

Another preservative people used was **spice**. In early Egypt, the bodies of kings were preserved with spices. When Marco Polo brought back spices from the East, people discovered that if they added spice to meat, it tasted better and would last longer. The spices could even mask the taste of meat that was not quite fresh. Today, we use many spices. 1

spice:
A plant substance added to food to preserve it or to flavor it.

1

Certain chemicals preserve the color, flavor, softness, or crispness of food. Preservatives have kept people from getting sick from food going bad. A chemical is added to bacon and cold cuts to preserve them. Without a preservative, what do you think would happen to the bologna sandwich in a lunch box from morning until noon? 2

2

Some additives that preserve food also help to make the flavor better. Sometimes, a lot of salt and sugar is added to packaged foods. Too much salt is bad for your health. Yet, people often eat ten times as much salt as they need, without ever realizing it.

People would be healthier if they ate no extra sugar. And they would also save their teeth from decay. There is enough natural sugar in fruits and vegetables for everyone's needs.

Check yourself

1. Name the gas that is used to make soda.
2. Why are food additives used?
3. What is a preservative?
4. Name a very old preservative.
5. In what ways do spices help food?
6. Name two preservatives that we eat much too often.

Coloring foods

How would you like to eat hot dogs that are gray in color? Yet, that is the natural color of hot dogs. When you buy oranges, do you choose bright orange ones? Would you buy margarine that was white?

Coloring is an additive used to make food look better to the buyer, or **consumer** (kən sōō′mər). Oranges look more orange when coloring is added. Yellow coloring is added to margarine to make it look like butter because consumers are used to spreading something yellow on bread. Red coloring is added to hot dogs, strawberry ice cream, fruit drinks, and many other packaged foods. 1

Some coloring is natural. It comes from brightly colored plant parts, such as berries. Other coloring additives are **artificial** (är′tə **fish**′əl), or made from chemicals.

consumer:
The person who buys and uses foods and other products.

artificial:
Not found in nature; made by people.

1

Processing foods

The foods people eat today are made mostly from the same or similar vegetables, fruits, and grains used as food long ago. But they look and taste very different. This is because some food may go through one or more **processes,** or changes, when it is prepared. The process may include grinding up the food and cooking it before packaging it. For

process:
To treat something so that it changes.

example, corn may be cooked, ground up, rolled out in sheets, dried, broken up, and baked. The result is corn flakes. 2

2

Some additives are used in the processing of food. Peanut butter you buy in stores today is creamy. 3 That is because an additive keeps the oil in the mix and prevents it from rising to the top. Would you like to eat ice cream that was full of tiny bits of ice? Ice cream has a thickener in it to make it smooth. Today, half of all the food people eat is processed. For example, many additives are used to make food smooth, light, crisp, or soft.

3

✓ Check yourself

1. What additives make food look better?
2. What is natural food coloring?
3. How is artificial coloring made?
4. What does it mean to process food?
5. How much of our food is processed?

137

Adding nutrients to food

Some additives make foods richer in **nutrients** (**noo'**trē ənts). When whole wheat is processed to make white bread, the nutrients that were in the brown part of the wheat are lost. Additives can put certain nutrients back into the white bread.

Often, nutrients must be added to make up for vitamins that were destroyed when the food was processed. The cooking process can destroy some vitamins. The extra nutrients **fortify** the food, or make it more healthful. It is possible to fortify even a glass of water or juice with vitamin drops. 1

nutrient:
The part of a food that is used by our bodies.

fortify:
To make more healthful.

1

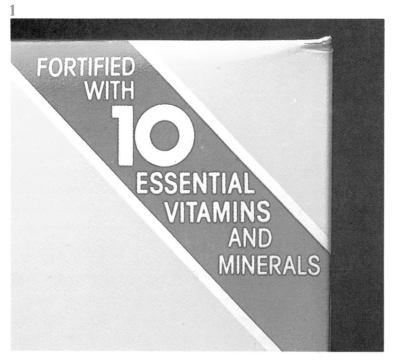

Both vitamins and minerals are added to enrich foods and give people what they need for good health. Several minerals, such as iron, may be added to cereals. Vitamin D is added to milk, and vitamin C is added to fruit drinks.

Activity

Experiment with additives

Try these activities at home or at school.

1. Compare the color of a green vegetable before and after it is cooked. Cook some with ½ teaspoon of baking soda. Compare.

2. Squeeze an orange and compare the taste of the juice with canned or frozen juice. 2

3. Prepare two cups with a small amount of fresh fruit in each—berries, peaches, bananas, or a mixture. Stir a teaspoon of sugar in one cup until all the fruit is coated. Put both cups in the refrigerator overnight. What happens?

2

In experiment 1, green vegetables are a brighter green when cooking starts, then they slowly become a duller green. Soda preserves the green color but destroys the vitamins.

In experiment 2, fresh-squeezed juice has a different flavor and is not as sweet as processed juice.

In experiment 3, the sugar preserves the fruit.

Check yourself

1. What is lost when food is processed?
2. How do some additives make food more healthful?
3. What vitamin is added to milk?
4. What is destroyed by cooking a green vegetable in soda?

Laws about additives

There are additives that no one wants to eat. Consumers have asked the government to pass laws to make sure that processed food is safe and healthful. There are laws that require white bread to be fortified. TV dinners must have a certain amount of nutrition. Coloring and flavoring that are not natural to the food must be listed as "artificial."

The law requires that all foods and additives in a package be listed on the label. They must be listed in the order of the amount in the package, from most to least.

 **Activity**

Reading food labels

1. Select six different packages of food. On most, you will find a place on the label where everything inside the package or container is listed. Everything is listed after the word "Ingredients."

2. Make a chart like the one shown. 1 Record information you find on each label by writing yes, no, and other words that describe the additives in the food you are checking.

1 **Additives**

	Salt	Sugar	Preservative	Coloring	Flavoring	Enriched/Fortified

Food

3. On the chart, put an *N* in the box if the color or flavor is natural. Put an *A* in the box if it is artificial.

Were you surprised at some of the strange additives listed? Did you know that so many foods had salt and sugar? Reading labels carefully will help you be a wise consumer.

✓ Check yourself

1. What has the government done to make sure that processed food is safe?

2. What must the label on a food package list?

3. Where is the ingredient present in the least amount listed on a food package?

Science in our world

Scientists test food additives on animals, such as rats and dogs, to help decide if they are safe.

Sometimes, an additive is found to cause cancer in animals. Then, the government will not allow it to be used. However, this is not always possible. For example, the additive that helps preserve certain meats is sodium nitrite. Some scientists fear that this chemical may cause cancer. But, the same additive keeps people from being poisoned by bad meat, so it is still used. Wise consumers should cut down on meats that contain sodium nitrite.

Careers

Food technologists (tĕk **nŏl′ə** jĭsts) investigate the composition of foods. They study the changes foods go through when they are stored. What they learn is used by processors and packagers.

1

142

Food technologists need a B.A. or a B.S. degree. Usually, they major in science or in food technology. They find positions with the government in the Food and Drug Administration. They also find jobs in the food and food-service industries. 1

Review the main ideas

In the United States, people have a better chance to be healthy today than at any time before in history. 2 Packaged foods stay fresh, and they are easy to prepare, good tasting, and enriched for good nutrition.

But sugar, salt, and some chemical additives are not good for you. The government can protect the consumer with testing and laws. But it is up to the consumer to read labels and decide which foods to buy and which to do without.

2

Chapter review

Science vocabulary

Choose the items in column B that best match the terms in column A.

A	B
1. additive	a. preserves and seasons food
2. preservative	b. to make more healthful
3. spice	c. user
4. consumer	d. to change
5. artificial	e. healthful
6. process	f. not natural
7. nutrients	g. used in processing food
8. fortify	h. keeps food from spoiling

What did you find out?

Answer the following.

1. The first food additive that kept meat and fish fresh was salt. (True/False)
2. Most people eat too much sugar. (True/False)
3. A preservative always makes a food taste better. (True/False)
4. Peanut butter has an additive that keeps it creamy. (True/False)
5. Spice cookies are fresher than other cookies. (True/False)
6. All additives are artificial. (True/False)
7. Most people eat too much salt. (True/False)
8. Vitamins are added to all packaged foods. (True/False)

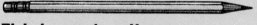

Think and write

1. Are fresh foods better for you than packaged foods? Why?
2. Name some food additives that are beneficial. Explain why they are used and how they are helpful.

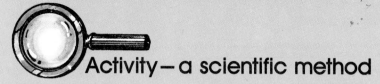

Activity—a scientific method

Purpose: to learn how some foods can be preserved.

Materials needed

8 plastic cups   ½ apple   knife   marking pen
½ banana   piece of lemon   water

Procedure

1. Use your marker to label each cup. Label two cups "plain," two cups "lemon," two cups "water," and two cups "lemon and water."
2. Fill each "water" and "lemon and water" cup half full with water.
3. Add several drops of lemon juice to each "lemon and water" cup.
4. Cut four slices of banana. Place a slice into one of each type of cup.
5. Squeeze lemon juice all over the slice in the "lemon" cup.
6. Repeat steps 4 and 5 using four apple slices.
7. Set the cups aside for 1 hour. Then, record what you see.
8. Allow the cups to stand overnight. Record what you see.

Observations

1. What happened after 1 hour? What happened the next day?
2. How do you explain what you saw?

2

Earth and space science

10. Minerals and rocks
11. Soil: a resource
12. Conserving water
13. Weather watch
14. Our solar system
15. New sources of energy

Minerals and rocks

In this chapter, you will find out that

- Minerals have special features that make them different from rocks.
- Minerals can be identified by their properties.
- Rocks are grouped as igneous, sedimentary, and metamorphic rocks.
- In the rock cycle, rocks change from one kind to another kind.

Before you begin

Have you ever seen a house built of rock? For centuries people have taken rocks from the ground. They have learned to shape the rocks, to carve into them, and to build with them. Buildings, homes, statues, and roads are made of rock. Many of these rocks come from places very far away from where they are used.

People take rocks from a place called a **quarry** (**kwôr′ē**). Rocks can be taken from quarries as blocks or in small pieces called trap rock. Trap rock is sometimes called "crushed stone." It is used in the building of roads. What is a rock? How are rocks formed? How can we put rocks into groups?

quarry:
The place from which rocks are removed.

149

What is a mineral?

There are more than 2,000 minerals. Some people think that rocks and minerals are the same thing. They are not.

Minerals have certain features. All minerals are found naturally. They are not manufactured, or made by people. Minerals are not formed from living things or the remains of living things. Coal is not a mineral because it was formed from plants.

Another feature of minerals is that they are solids. Oil is not a mineral. Oil is formed from the remains of plants and animals. Also, oil is a liquid.

Do you think steel is a mineral? Before you answer, ask yourself these questions. Is steel natural or manufactured? 1   Is it formed from living things? Is steel a solid? Steel is not a mineral because it is manufactured.

Minerals are always made of the same **elements** (ĕl'ə mənts). Some minerals are made of only one element. Gold and iron are minerals made of only one element. Most minerals are made of different elements. The elements are combined in a special way in each mineral.

element:
A substance that cannot be broken down by chemical means.

1

Another feature of minerals is that the **atoms** (ăt′əmz) that form them are always arranged in the same way. They form crystals. A crystal is a solid with definite edges, corners, and angles. Crystals of the same mineral can be different sizes. But they are always the same shape. Salt is the mineral called halite (hăl′īt′). Halite crystals are always cubes. 2

2

Ores and gems

People have always used minerals. Some minerals, such as gold and silver, are used to make jewelry. Aluminum is used to make cans. Iron is used to make steel for building. Many different minerals are used for many purposes.

Some minerals are **ores**. An ore contains a lot of a metal or a nonmetal. The mineral hematite (hěm′ə tīt) is an ore. It contains a lot of iron. But it also has other substances in it. The iron can be taken out of the ore by heating it to a very high temperature.

Some minerals are rare and valuable. They are also long lasting and beautiful. These minerals are called **gems**. 3 Diamonds, rubies, and emeralds are gems. Some gems, such as opals, are not as rare or valuable as others. Birthstones are gems. What is your birthstone?

3

✔ Check yourself

1. What are five features of minerals?
2. What is an ore?
3. What is a gem?

Identifying minerals

How can we tell minerals apart? The **properties** (**prŏp′**ər tēz) of minerals help identify them. We can perform tests on a mineral to find out what properties it has.

Color is a property of a mineral. Minerals have many different colors. Color is easy to observe. But color can fool you. Many minerals have the same color. For example, iron pyrite (**pī′**rīt) is yellow in color.  It looks like gold. Many people thought it was gold! And pyrite became known as "fool's gold."

Some minerals are always the same color. Some minerals are found in different colors. They sometimes have different names. Quartz (kwôrts) is clear, purple, or yellow. Purple quartz is called amethyst (**ăm′**ə thĭst).  Yellow quartz is called citrine (**sĭ′**trĭn). Using color to identify a mineral can be helpful. It can also be confusing!

You can also look at a mineral's **streak**. Streak is the color of the powdered form of a mineral.

properties:
Traits that help tell one thing from another.

streak:
The color of the powdered form of a mineral.

Activity

Mineral streak test

1. You will need a streak plate (a piece of unglazed porcelain or ceramic tile), pyrite, quartz, and three to six other mineral samples.

2. Rub the pyrite across the plate to obtain a streak.

3. What happened? Is the color of the streak the same as the color of the mineral?

4. Streak the plate with quartz. What happened?

5. Obtain the streak of other minerals. What happened?

6. Compare the color of the streak of each mineral with the color of the mineral. Make a chart. 3

3

Pyrite produces a black streak when it is rubbed across the plate. But, pyrite is yellow in color. The color of the streak and the mineral may not be the same. Some minerals do not leave a streak. Quartz does not leave a streak. This is because it is harder than the plate.

Hardness is another property of minerals. In 1822, a mineral hardness scale was developed by Fredrick Mohs (mōz). 4 It is used to compare the hardness of an unknown mineral with the ten minerals on the scale.

4
MOHS HARDNESS SCALE

Softest	1	Talc
	2	Gypsum
	3	Calcite
	4	Fluorite
	5	Apatite
	6	Orthoclase
	7	Quartz
	8	Topaz
	9	Corundum
Hardest	10	Diamond

Each mineral on the scale can scratch any other mineral with a lower number. A harder mineral will always scratch a softer one. Look at the hardness scale. Which minerals will scratch quartz? Which mineral is the hardest?

Common objects can also be used to compare hardness. Try scratching a mineral with your fingernail and with a steel knife. See if the mineral will scratch steel or a piece of glass.

Minerals also shine or reflect light in a certain way. This is called **luster**. Some minerals shine like polished metal. 1 Other minerals are dull. They do not look like metal. There are many different lusters. Quartz has a glassy luster. Talc has a pearly luster. 2 Gypsum has a silky luster.

luster:
The way a mineral looks when it reflects light.

1

2

✓ Check yourself

1. Why is color not a good way to identify a mineral?

2. Are the color of a mineral and its streak always the same?

3. How is the Mohs Hardness Scale used to identify a mineral?

4. What is the way a mineral reflects light called?

Rocks

All rocks are made of minerals. Minerals are the "building blocks" of rocks. In one rock, you can usually find several different minerals.

There are many different kinds of rocks. But **geologists** (jē ŏl'ə jĭsts) put rocks into three groups. The way a rock is formed tells us to which group, or class, the rock belongs.

One class of rock is **igneous** (ĭg'nē əs) **rock**. Igneous rocks are "fire-formed" rocks. They are formed from **magma** (măg'mə). Magma is very hot, liquid rock. When magma cools, igneous rocks form. Some igneous rocks, like granite (grăn'ĭt), have large crystals of minerals. 3 Some have

geologist:
Scientist who studies the structure of the Earth.

igneous rock:
Rock formed from melted rock that cools.

magma:
Hot, liquid rock beneath the Earth's surface.

3

155

1

2

small crystals. Some rocks look like glass. In fact, obsidian (ŏb **sĭd′**ē ən) is sometimes called volcanic glass. 1

Some igneous rocks, such as pumice (**pŭm′ĭs**), have many holes. Pumice looks like a sponge. Sometimes it is so light, it can float in water! 2

Another class of rock is **sedimentary** (sĕd′ə **mĕn′**tə rē) **rock**. These rocks are made of small pieces of other rocks. These pieces are pressed and cemented together. The pieces of rocks may be different sizes. Sand, clay, gravel, and pebbles are parts of some sedimentary rocks. Even pieces of animal shells may be part of these rocks.

sedimentary rock: Rock formed from sediments or pieces of other rock.

Sandstone and limestone are sedimentary rocks. Limestone is made from the shells of animals. 3 Can you guess what kinds of particles make up sandstone? 4

3

4

The third class of rocks is **metamorphic** (mĕt′ə môr′fĭk) **rock**. These rocks are sometimes called "changed rocks." Metamorphic rocks are formed deep inside the Earth. When other kinds of rocks are buried deep in the Earth, they may change. Pressure and high temperatures change the rocks to different rocks. For example, the igneous rock granite can be changed to gneiss (nīs). 5 The sedimentary rock limestone can be changed to marble. 6 In fact, most metamorphic rocks are formed from sedimentary rock.

metamorphic rock: Rock changed by pressure and high temperature.

5

6

157

Activity

Identify rocks in buildings

1. You need at least ten photographs of different buildings made of rock: houses, schools, libraries, monuments, and churches.

2. Mount the photos on poster board or oak tag.

3. Identify the type of rock seen in each photograph. Some rocks commonly used in buildings are granite, sandstone, limestone, marble, and slate. Write this information in the form of a caption next to each picture.

The rock cycle

Rocks can be changed from one type to another. On the surface of the Earth, all rocks are broken apart by running water, wind, snow, and ice. Even heat can break apart rocks. When rocks are broken apart this way, we say they have **weathered**. Weathering can also change the minerals in rocks.

Sedimentary rocks are formed from pieces of metamorphic and igneous rocks. Even pieces of other sedimentary rocks can be part of another sedimentary rock. Metamorphic rocks can be changed to other kinds of rocks, too. So can igneous rocks!

weathering: Breaking apart and breaking down of rocks on the surface of the Earth.

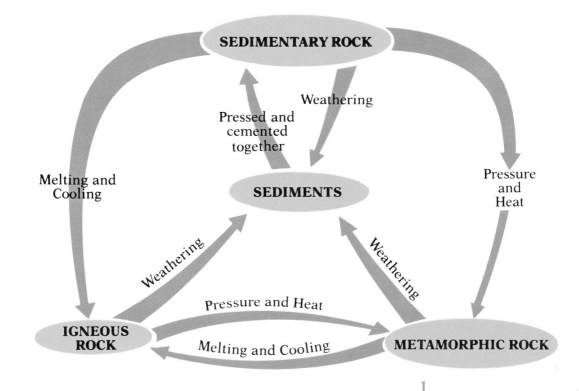

SEDIMENTARY ROCK

Weathering

Pressed and cemented together

SEDIMENTS

Melting and Cooling

Pressure and Heat

Weathering

Weathering

IGNEOUS ROCK

Pressure and Heat

Melting and Cooling

METAMORPHIC ROCK

1

The changes in rocks from one kind to another make up the **rock cycle**. 1 The rock cycle shows the many different ways that rocks change. Can you name some ways that rocks are changed? The rock cycle also shows how the different types of rocks are related to each other.

rock cycle:
The changes in rocks from one kind to another kind in a continuous cycle.

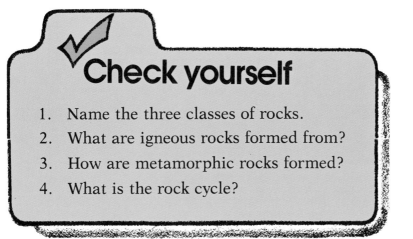

Check yourself

1. Name the three classes of rocks.
2. What are igneous rocks formed from?
3. How are metamorphic rocks formed?
4. What is the rock cycle?

Science in our world

In many places, rocks have been used to build walls. Rocks were piled one on another to make a wall. These walls lasted a long, long time and are still in use. Rock walls can be seen in parts of Europe, New England, and the hill country of Texas. Even after steel wire was invented, rocks were still used for fence posts. In central Kansas, for example, there was not much wood, and columns of rock were used instead.

Careers

Sometimes, certain buildings are decorated with quarried rocks, such as marble and granite. Such decoration is beautiful, but very expensive. The crafts people who work on these rock coverings, or

facings, are called **stone masons** (stōn mā′sənz). Usually, they work from a set of drawings, or blueprints, that indicate where each rock should be placed. Sometimes, the rocks are put on walls in patterns. Often, masons cut rocks into special shapes and sizes.

Stone masons must be good at using hand tools. They use trowels, hammers, and chisels. They also use power tools to cut rock and to smooth the surface of a rough rock.

Review the main ideas

Minerals are found in nature. They are solid and are not formed from the parts of living things. Each mineral is always made of the same element or elements. Its atoms are always arranged the same way. Gems are rare and beautiful minerals. Minerals can be identified by their properties. Some properties we can use are color, streak, hardness, and luster.

Rocks are mixtures of minerals. Igneous rocks form from magma, or lava that cools. Sedimentary rocks are formed from weathered particles of other rocks and minerals. When rocks are changed by pressure and temperature, metamorphic rocks are formed.

The rock cycle shows us how the three groups of rocks are related and how rocks can change from one type to another. Rocks can be changed by weathering, pressure and heat, and by melting completely and cooling again.

Chapter review

Science vocabulary

Choose the item in column B that best matches the term in column A.

1. granite
2. quarry
3. ore
4. marble
5. sandstone
6. limestone
7. rock

a. mixture of minerals
b. formed from animal shells
c. formed from weathered sediments
d. place from which certain rocks are taken
e. an igneous rock
f. mineral that contains usable amounts of a metal
g. a metamorphic rock

What did you find out?

Answer the following.

1. The group of rocks formed by pressure and heat is called _____ .

2. The change that rain, water, and wind cause in a mineral or rock is called _____ .

3. The softest mineral on the hardness scale is _____ .

4. The hardest mineral is _____ .

5. The group of rocks formed when magma cools is called _____ .

6. A mineral is always made up of the same _____ .

7. The powder of a mineral left on a plate is called its _____ .

8. A mineral that looks like polished metal has a metallic _____ .

Think and write

1. How are the three groups of rocks related to each other?

2. Which group of rocks do you think would be most likely to contain fossils? Which group would be the least likely to contain fossils? Explain your answer.

3. How do you think an igneous rock could become part of a sedimentary rock?

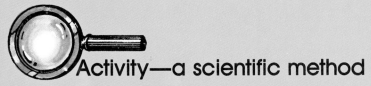

Activity—a scientific method

Purpose: to examine samples of rocks and record their characteristics.

Materials needed

granite, sandstone, limestone, marble, slate, and a hand lens

Procedure

1. Use the hand lens to look at each sample carefully. Note the color(s) and texture of each rock.

2. Look carefully for particles of minerals in each rock. Look carefully for different layers.

3. Make a chart of your observations.

Observations

1. In which rocks could you see sediment particles? In which rocks could you see mineral particles?

2. In which rocks could you see layers?

3. Arrange the rocks in order from the one with the largest particles to the one with the smallest.

11

Soil: a resource

In this chapter, you will find out that

- Food-growing lands are a natural resource.
- The greatest danger to soil comes from erosion.
- Soil conservation keeps land productive.
- Proper farming methods help to conserve soil.

Before you begin

Count slowly from 1 to 60. When you finish, about 1 minute of time will have gone by. That is not very long, is it? While you were counting, about 172 babies were born. In the next minute, another 172 babies will be born.

Today, you share the world with about 4 billion other people. If this birth rate continues, there will be about 6.5 billion of us in the year 2000. All of us must learn to share our Earth's valuable natural resources with each other.

Things that we value

A resource (rē′sôrs′) is something people consider valuable. When we speak of **natural resources,** we mean materials found in nature that are valuable to us. Trees are a natural resource. So are metals such as iron, lead, and copper. Fossil fuels such as oil, coal, and natural gas are also natural resources. They are valuable to us for many different reasons. We use natural resources to make tools, to build buildings, and to manufacture all kinds of goods. But there is one natural resource that we sometimes forget about. We see it everyday. It is soil. 1

natural resource: Material that is found in nature and is of value to humans.

1

Soil is one of our most valuable natural resources. It covers the surface of the Earth. In fact, soil is often called "earth." Soil is made up of bits of rock and contains organic (ôr găn′ĭk) materials. These are made by the animals and plants that grow and die in it. Without good soil, there would be no green plants, no wildlife, and no food.

Our food-producing lands

Every day, Earth's farmlands produce about 1 kilogram (about 2¼ pounds) of grain for every human being. 2   But all parts of the world do not produce the same amount of grain. Some countries produce more food crops than others. There are many reasons for this. Some soil is not suitable for growing crops. Or there may not be enough water

2

available. In some places, the weather is too cold or too dry and hot for growing crops.

If soil is fertile (**fûr′tl**), or rich in minerals, and there is enough water and sun, crops will grow well. But soil is not a simple thing. Have you ever held some soil in your hand? It may be difficult to believe, but the soil you held may have taken more than 1000 years to form! How is soil formed?

Over many years, rock is worn down. Plants begin to grow in the bits of rock. Some animals feed on the plants. The plant roots break down the rock bits even more. The plants and animals die, and their remains decay. Thus soil is formed.

Crops grow in the upper layer of soil, called **topsoil.** 3 Farming is best where the topsoil is rich and a good supply of water is available. But care must be taken to protect the soil. **Erosion** (ĭ **rō′zhən**) is the greatest danger to soil. Winds blow away the topsoil. Water washes it down hills and into rivers. The rivers carry it away. Erosion can change rich, food-growing soil into a desert. 4

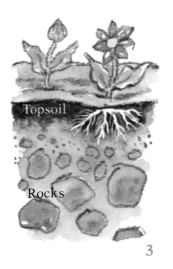

3

topsoil:
The fertile top layer of soil in which plants can grow.

erosion:
Carrying away of topsoil by wind and water.

4

Every year, about 4.5 billion metric tons of topsoil are lost due to erosion. There are several ways to protect cropland from erosion. Protecting the soil from erosion and keeping it healthy is called soil **conservation** (kŏn′sər **vā′shən**).

conservation:
Saving a resource.

✔ Check yourself

1. List three things that are considered to be natural resources.
2. Name two things which make up soil.
3. What causes different parts of the world to produce different amounts of grain?
4. Define "topsoil."
5. How does erosion hurt topsoil?
6. How does conservation help topsoil?

Soil conservation

Have you ever been in the country during a rain storm? If you walk near a stream after a heavy rain, you may see water that is muddy, like this. 1 Where did the mud come from? You can find the answer without going to the country. Try this simple activity.

1

Activity

Where does mud come from?

1. You need some old newspaper, 2 aluminum pans, 2 clear glass bowls, some soil, a watering can, and some grass clippings and leaves.

2. First, place newspaper around your work area. Then, fill both pans halfway with soil.

3. To one pan, add grass clippings and leaves. Pack these down on top of and around the soil.

4. Prop up one end of each pan with a stack of books, or a wood block, or anything similar. The high ends of each pan should be equal in height. That way, water can run down each pan at about the same angle.

5. Place a glass bowl at the lower end of each pan to catch the water.

6. Fill the watering can, and sprinkle water on each pan, as if it were raining. **2** The water should run down the pans and collect in the bowls.

7. Compare the water in both bowls. Is there a difference? Why or why not?

The color of the water in each bowl was probably different. The ground cover of leaves and grass protected most of the soil in one pan. But the bare soil in the other pan was not protected. Some of

1

that soil washed away, or eroded, and made the water in the bowl muddy. 1

The activity shows one way to stop erosion. Soil will not wash away easily if it is covered by plants. Plants protect the soil from erosion.

Protecting soil from erosion

Farming the land can increase the chance of erosion. Therefore, it is important to plant crops so that wind and rain do not carry off the good topsoil. Here, the land is rolling, or slightly hilly. 2 If the farmer had planted the potatoes in straight rows up and down, what would probably have happened? Instead, the potatoes are planted to follow the shape of the land. This method of farming is called **contour** (**kŏn′**to͝or) farming. Contour farming lessens the chances of soil erosion.

Another simple but very successful way of conserving soil is called **terrace** (**tĕr′**əs) farming. 3

contour:
Following the outline of the land.

terrace:
A raised bank of earth with a flat top.

2

3

Terracing is usually done on farmland that is quite hilly or steep. The land is planted in steps to slow down the flow of water. This keeps the water from carrying away the topsoil.

Check yourself

1. What type of ground cover protects soil from erosion?

2. Name the method of farming that follows the outline of the land.

3. What do we call a raised bank of earth with a flat top?

4. How does terracing reduce erosion?

Other conservation problems

The soil must be protected from things besides erosion. For example, suppose the same crop is planted year after year. Each year, this crop takes the same nutrients from the soil. Soon, the supply of these nutrients is used up. We say the soil is worn out. Different crops take different nutrients from the soil. By growing one crop the first year and a different one the following year, farmers give the soil a chance to rebuild itself. This method of farming is called **crop rotation** (rō tā′shən).

The animals that provide us with food can destroy the topsoil, too. Cattle and sheep graze, or feed, on grass. 1 Overgrazing on the same piece of

crop rotation:
The practice of not planting the same crop in the same field year after year.

1

land causes the grass to thin out. Wind and water begin to carry away the topsoil.

The forests and woodlands that surround our food-producing lands and grasslands must be protected, also. Where forests are cut down, the soil erodes. Trees and other forest plants normally hold

172

moisture in the ground. When they are cut down, they can no longer do that job. Nothing is left to catch the rain water.

Activity

Renewable resources

You have learned something about soil, forests, and grasslands, and how important these natural resources are to humans.

1. Use the information from this chapter and from library and reference books to complete a chart like the one below. 2

2. When your research is finished, mount your chart on construction paper.

3. If you want, illustrate the chart with drawings or photographs that describe your research.

RENEWABLE RESOURCES		
Resource	How it is managed	Problems
1. Fertile soil 2. Forests 3. Grasslands		

Everyone's chart will look different. In one class, students found out about the farming and cutting methods used to help prevent soil erosion and the loss of forests and woodlands.

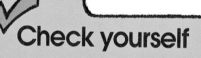

Check yourself

1. What happens to soil if the same crop is planted in it every year?
2. Why is crop rotation a good idea?
3. What do wind and water do to topsoil?
4. What happens to soil when forests are cut down?
5. What does overgrazing do to grass?

Science in our world

Early in our history, some people began to see the land as a valuable natural resource. In 1681, William Penn signed a law that helped conserve our forest lands. The law required that when a forest was cut, every fifth acre would be left uncut. In 1933, the Soil Conservation Service was established. Early work in soil conservation was done under the direction of this service. International Soil Science Societies meet in different parts of the world. Today, more and more people are coming to know that the strength of a country depends, in part, on the health of its soil.

Careers

Often, farmers and ranchers need assistance in dealing with soil erosion. **Soil conservationists** are the people they call upon for help. 1 Soil conservationists look for the source of the erosion problem and suggest ways to combat it. If erosion is caused by water running off a hilly land, they may recommend terracing. If wind is causing soil erosion, they may recommend growing hedges that act as windbreaks. Usually, they plan soil conservation programs for entire communities.

Soil conservationists must have a college degree. Usually, their fields of study are agriculture and related sciences.

1

Review the main ideas

Earth's natural resources must be shared by more and more people. Soil is one of our most precious natural resources. It is made up of bits of rock and materials from plants and animals. Not all soil is fertile, or suitable for growing crops. This is one reason why some countries produce more food crops than others.

Crops grow in the upper layer of soil called topsoil. Topsoil must be protected from erosion, which is the wearing away of soil by water and wind. Proper farming and grazing methods can help prevent erosion. Some of these methods include crop rotation, contour farming, and terracing. Care in cutting down forest areas and in grazing cattle and sheep also helps to protect soil.

Chapter review

Science vocabulary

Choose the items in column B that best match the terms in column A.

A	B
1. natural resource	a. effect of wind and water on soil
2. erosion	b. planting different crops each year
3. topsoil	c. valuable material from nature
4. graze	d. to feed on vegetation
5. crop rotation	e. rich layer of soil in which crops are grown

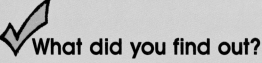

What did you find out?

Answer the following.

1. Wind and water can cause the _____ of precious soil.
2. Planting crops to follow the shape of the land is _____ .
3. Very hilly or steep farmland can best be planted by using a method called _____ .
4. Crop rotation gives worn out soil a chance to _____ itself.
5. Trees and other forest plants help prevent erosion because they (hold back/speed up) the flow of water.

Think and write

1. What is meant by the term "natural resource"? Name three.
2. Describe two farming methods that help prevent soil erosion.
3. Why are forests and woodlands that surround crop-growing areas important?

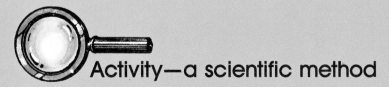

Activity—a scientific method

Purpose: to compare how seeds grow in five kinds of soil.

Materials needed

5 plastic cups   10 pea seeds   gravel   sand   potting soil
vermiculite   pencil   medicine cup   water   ruler

Procedure

1. Fill one cup ¾ full of sand. Fill a second cup ¾ full of vermiculite. Fill a cup ¾ full of gravel. Fill a cup ¾ full of potting soil. Fill a cup ¾ full with a mixture. Combine 1 part vermiculite, 1 part sand, and 1 part soil. Soak ten pea seeds overnight.
2. In each planting medium, make a hole about 1½ centimeters deep. Drop two peas into each hole. Push them down with the pencil's eraser.
3. Gently cover each hole by patting down the soil, sand, etc.
4. Add two medicine cups of water to each cup. Put the cups in a corner that does not get direct sunlight. Check the cups each day. If they are dry, add two medicine cups of water.
5. As soon as a plant comes up, move it into the sunlight.
6. Measure each plant each day. Continue to do this for 2 weeks.
7. Record the day each plant comes up. Record when you have to add water to each cup and the growth of each plant.

Observations

1. In what order did the plants appear?
2. Which plant grows best? Which plant grows poorest? Why?
3. Why is soil considered a valuable resource?

12

Conserving water

In this chapter, you will find out that

- ■ The Earth's water comes from two sources.
- ■ There are two major problems with our water supply.
- ■ Water is carried to cities by ducts, or pipes.
- ■ There are ways to clean and conserve water.
- ■ The water on Earth today is the same water that was here when life began.

Before you begin

Have you ever thought about what makes up your body? Does it surprise you that your body is about two-thirds water? All living things are made up mostly of water. It keeps your body working properly.

Plants and animals need clean water to remain healthy. Pure water is necessary to support human, animal, and plant life. It is a life-giving natural resource.

There is no new water on the Earth. The same water is used over and over again. That has been true since there was first life on this planet. How can water be used over and over? Can used water be made clean enough to drink?

Do you know how drinking water comes to your home? Do you know what happens to it before it reaches you?

Two sources of water

Where does water come from? If you look at a globe of the world, you will notice that much of Earth's surface is covered by water. Water that can be seen above the ground is called **surface (sûr′fəs) water.** Oceans, lakes, rivers, ponds, and streams are all examples of surface water. 1   Surface water is one of the major sources of water in the world.

The other major source of water lies beneath the surface of the Earth. It is called groundwater. It may rise to the surface when it bubbles up in a spring. Or mechanical means may be used to bring groundwater to the surface. Pumps and wells are often used for this purpose. 2

surface water:
Water that can be
seen above ground.

1

2

The water cycle

The amount of water in the world does not change. When the sun shines, it produces heat. This heat causes water to dry up, or evaporate, from bodies of water into the air. The moist air rises and cools. The moisture in the air condenses, or turns back to liquid. It forms tiny water droplets or ice crystals. Clouds are formed. Some of the cloud particles bump into each other and join to grow in size. On other particles, more moisture condenses so they grow larger. When they reach a certain size, the particles fall as rain or snow. Rain and snow are forms of **precipitation** (prĭ sĭp′ĭ tā′shən). The diagram shows the process that is known as the water cycle. 3

This cycle is repeated over and over. A cycle is any series of steps that is repeated. Through the water cycle, water is returned to Earth to be used again and again.

precipitation:
Water that falls from the air to the Earth. It may fall as rain, snow, sleet, or hail.

3

Water vapor condenses to form clouds

Winds move the clouds

Water falls as precipitation

Water vapor rises

Water runs off

Water evaporates

Ocean

Problems with water

In the past, when there were fewer people, each family or tribe got its own water from streams or springs. As the world's population grew, the need for clean water increased. People, animals, and plants cannot live without water. Yet, while the amount of water in the world does not change, the number of people in the world keeps growing larger. There are many people living in large cities. It is very hard to bring enough fresh water to large numbers of people living in cities. This is one of the major problems about water that we face today. More and more people need water—yet, there is only a fixed amount of water in the world.

A second problem is that water must be clean and pure if it is to help support life. Yet, for many reasons, water is becoming more and more polluted. Polluted water contains substances that are harmful, and even deadly, to living things. **1**

Water becomes polluted in many ways. First, there is natural pollution. Soil, leaves, and tiny organisms get into water from nearby land. Then, there is pollution caused by human activity. Coal, oil, and gas fuels give off chemicals when burned. **2** Rain containing these chemicals falls as acid rain, which is harmful to many forms of life. Harmful factory chemical wastes and sewage also pollute water. **3**

1

2

3

✓ Check yourself

1. Define "surface water."
2. What do we call water that falls through the air?
3. How is groundwater brought to the surface?
4. Describe the water cycle.
5. What water problems does population growth cause?
6. What do we call water that is not clean and pure?
7. How does nature cause water pollution?

City water supplies

Hundreds of years ago, people lived near sources of water. As cities grew larger, it became harder to supply enough fresh water. Water had to be carried to cities from sources farther away. Now, many cities use huge pipes, or ducts, to carry water. 1

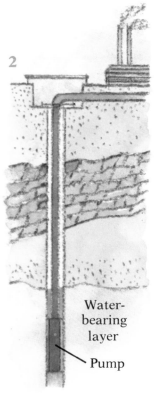

Water-bearing layer

Pump

Many cities store fresh water in reservoirs above the ground. Other cities use deep wells for water supplies. 2 As the population grows, more wells have to be drilled. A network of pipes carries the underground water to the cities.

Salt water is not suitable for drinking. Some cities build plants to remove salt from ocean water, turning it into fresh water. This process is called desalting. 3   It is very expensive.

3

How to clean water

There are a number of different ways in which water can be cleaned and made fit for use. One system imitates the water cycle. In this process, water is heated so that it becomes vapor, or steam. As it evaporates, the water becomes pure. Then, the vapor is chilled and turned back into liquid water. This process is called distillation (dĭs'tə lā'shən). 4 The pure water formed by this process is called distilled water.

A second system of cleaning water is called filtration (fĭl trā'shən). In this process, water is poured through layers of loose material, such as sand or charcoal. Sand or charcoal acts as a strainer, trapping impurities that are in the water.

4

Pure water

Salt water

**SIMPLE
DISTILLATION**

You can try to filter water yourself. It is not the same as the huge water-treatment plant that may be in your town, but the idea is the same.

Activity

Filter water

1. You need a coffee can of muddy water, coffee filters, and a clear glass jar. You also need a funnel that fits over the jar.

2. Mix mud and water inside the coffee can.

3. Place the filter paper inside the funnel.

4. Place the funnel on top of the jar.

5. Pour the contents of the coffee can into the funnel, a little at a time.

6. When water starts dripping into the jar, examine it. How does it compare to the water in the coffee can?

7. Does the water stop going through after a while? Why?

8. After you have some water in the jar, start with new filter paper and run it through again. What happens?

If the water stopped, the grains in the mud were large and started to clog up the filter paper. By filtering it several times, the water can become quite clean. This is similar to what happens when large amounts of water are filtered. Then, if the

1

water is still cloudy, some chemicals are used to remove any remaining particles. 1

Chlorine is added to the water in pools to kill harmful microscopic living things. Chlorine is also added to water supplies for the same reason, but in much smaller amounts. One drop of chlorine is added to a million drops of water. Chlorine kills the living things that could cause diseases such as typhoid, cholera, or dysentery.

Check yourself

1. How do cities carry water from sources that are far away?
2. What process removes salt from ocean water?
3. What process cleans water by straining?
4. Name a chemical that is used to clean water.

Is clear water always pure?

Many impurities in water are not visible to the naked eye. You can see that water containing mud or leaves is polluted. But can you see that water containing microorganisms is polluted?

Even water that looks absolutely clear and transparent may have bacteria, minerals, and organisms in it. It is the minerals in water that give it its taste. Distilled water, which has no minerals in it, has no taste. Some people complain that it tastes flat.

There are certain places in the world that are known for their mineral water. 1 Some people go to great lengths and expense to drink it.

1

Conserving water

Do you have any idea how much water we use? According to some experts, the average person in the United States uses about 400 liters (100 gallons) every single day! In a large city, such as New York City, about 8 billion liters (2 billion gallons) are used on just one hot summer day. That is enough water to fill about 20,000 Olympic-size swimming pools.

Although water is renewed through the water cycle, sometimes there are shortages. During a water shortage, people try to conserve water. 2 How careful are you in your regular use of water?

2

Activity

How well do you conserve water?

There are lots of ways you can cut down on the water you use. Here are some you can do easily.

1. Read the chart below.

Activity	Water Conservation Method	Amount of Water Saved	
		Liters	Gallons
Brushing your teeth	Use a glassful instead of running the faucet	12	3
Getting a drink of water	Put a jar of water in the refrigerator instead of running water to get it cold	4	1
Taking a shower	Limit yourself to 3-5 minutes	16-24	4-6
Taking a bath	Only fill the tub halfway	40	10
Flushing the toilet	Only flush when necessary	28	7
Washing dishes	Fill the sink, using the stopper; do not run the water continuously	28	7

2. On a piece of paper, write down the amount of water you saved today for each activity.

3. Add up the total amount of water saved.

How well did you do? Rate yourself this way: 0 to 40 liters saved = water consumer; 41 to 80 liters = getting better; 81 to 120 liters = you are on the right track; 121 to 160 = a real water miser.

189

✓ Check yourself

1. Name three impurities that may be in clear water.

2. What do people try to do during a water shortage?

3. How many gallons of water can you save by limiting your showers to 3–5 minutes?

Science in our world

Sometimes, we do not realize how important something is until it is almost gone. 1 Water is a precious resource. We should think about how much we are using when we use it. Do not take water for granted. Can you imagine what it would be like if you turned on the tap and nothing happened? Or, if the government gave you a fine for watering your lawn, or having a leaky water pipe? These things are not impossible. They have happened in places where there were water shortages.

But it does not take a shortage to realize that as the population grows, more water is needed. Cleaning and reusing water is the only way that people will have enough water in the future.

1

190

Careers

As the need for water increases, more water treatment plants are being built to provide people with pure water. These plants employ many people. One important job is that of a **waste-water treatment plant operator.** 2 These employees operate the equipment that removes waste materials from water. They check instruments that measure the purity of the water. At different times, they take samples to make sure that the treated water is safe for people to use.

Waste-water treatment plant operators usually have a high school diploma and receive training provided at the plant where they work. They are usually good at mathematics and enjoy working with technical equipment.

2

Review the main ideas

All living things are made up mostly of water. The amount of water in the world does not change.

Water at or near ground level is called surface water. There is also water under the ground. Surface water and underground water are both the result of the water cycle. Water evaporates as it warms. It rises as water vapor into the air. The vapor cools and forms clouds. When precipitation from the clouds falls to Earth, the cycle begins again.

As cities grow larger, more and more water will be needed. Often, the water for cities must be brought from great distances. Some cities treat ocean water to turn it into fresh water.

There are a number of different ways to clean water. Three of the most important ways are distillation, filtration, and chlorination.

Chapter review

Science vocabulary

Choose the items in column B that best match the terms in column A.

A

1. surface water
2. evaporate
3. condense
4. precipitation
5. water cycle
6. desalting
7. distillation
8. filtration
9. chlorine

B

a. become a liquid
b. a method of cleaning water that is most like the water cycle
c. chemical sometimes added to water
d. process of removing salt
e. rain and snow
f. evaporation and condensation of water
g. lakes, ponds, streams
h. pouring water through layers that strain it
i. dry up

What did you find out?

Answer the following.

1. Two major sources of water are surface water and water that is _____ .

2. The moisture in the air _____ to form tiny water droplets or ice crystals.

3. Rain and snow are forms of _____ .

4. Because of human activity, some of our waters have become _____ .

5. Heating water until it forms vapor, then chilling the vapor and trapping the pure water, is the process of _____ .

Think and write

1. List the stages of the water cycle. Make a chart showing them.
2. List four ways that are used to clean water. Describe two.
3. Design a poster or an advertisement describing at least three ways people can practice water conservation.

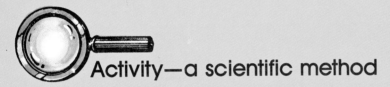

Activity—a scientific method

Purpose: to purify water by using chemicals.

Materials needed

muddy water

transparent container

aluminum sulfate solution

calcium hydroxide solution

1 teaspoon

Procedure

1. Get a container of muddy water.
2. Add a teaspoon each of aluminum sulfate and calcium hydroxide. Stir.
3. Watch the container for a few minutes. Record what you see.
4. Check the water from time to time throughout the day

Observations

1. Does the water get as clear as drinking water?
2. What is on the bottom of the container?
3. How did it get there?

13

Weather watch

In this chapter, you will find out that

- Moving air masses bring weather changes.
- Temperature, humidity, and air pressure affect the weather.
- There are different kinds of clouds.
- Meteorologists gather weather data and make predictions.

Before you begin

How often do you depend on the weather report? You want to know when to carry an umbrella or wear a jacket. The weather report may not always be exactly right, but it can be close.

The information you get on radio and TV has come from meteorologists. They are scientists who study weather patterns. Using these patterns, meteorologists can predict weather conditions. What tools does a meteorologist use?

Weather changes do not just suddenly happen. Denver's weather today may be Chicago's weather tomorrow. It may also be Baltimore's weather the day after tomorrow. Did you know that in the United States weather generally moves from west to east? Why do you suppose that happens?

Do you know what an air mass is? Did you know that moist air is lighter than dry air? Do you know how temperature, humidity, and air pressure affect the weather?

How air moves

Meteorologists study how air travels. Air travels in large volumes called masses. Sometimes, an **air mass** is warm. Sometimes, it is cold. This is because air takes on the temperature of the land or water it is near. When a mass of hot air hits a mass of cold air, the weather changes.

The line where one air mass meets a different air mass is called a **front** (frŭnt). In the United States, air masses usually move from west to east because of the Earth's rotation.

Look at the first picture. 1 It shows a fast-moving warm air mass that has bumped into a slower-moving cold air mass. The faster-moving warm air mass gradually rises over the cold air mass. This happens because warm air is lighter than cold air. The edge between the two air masses is the front. The front is named after the faster-moving mass of air. As the warm air rises, clouds form at the edge of the warm front. The clouds are close to the ground, and rain falls from them.

Now look at the next picture. 2 Which air mass is moving toward the east at a faster speed?

air mass:
A large quantity of the same type of air.

front:
The line where one air mass meets a different air mass.

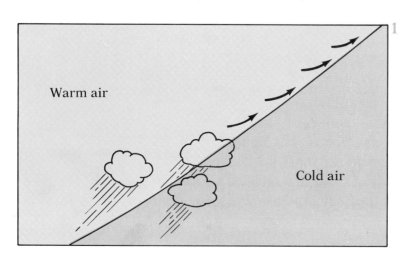

Warm air

Cold air

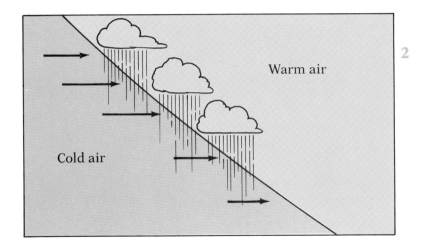

The fast-moving cold air mass pushes under the slower-moving warm air mass. This happens because cold air is heavier than warm air. The cold air mass acts like a wedge. Clouds form as the warm air is pushed upwards. The clouds formed are usually higher than those in a warm front. Often, there will be heavy rain.

Temperature

Do you know what the temperature is right now? 3 The hotness or coldness of the air is called the temperature. It is measured with an instrument called a thermometer. Temperature is a very important element in weather. People are strongly affected by its changes. Which do you like best: a hot, a warm, a cold, or an icy day?

Activity

Comparing temperatures

CITY OF FAIRHAVEN HOURLY TEMPERATURE READINGS (Degrees Celsius)			
May 25		**May 26**	
4 A.M.	21	4 A.M.	22
5 A.M.	21	5 A.M.	22
6 A.M.	21	6 A.M.	21
7 A.M.	20	7 A.M.	21
8 A.M.	22	8 A.M.	22
9 A.M.	23	9 A.M.	23
10 A.M.	25	10 A.M.	24
11 A.M.	26	11 A.M.	25
12 noon	26	12 noon	26
1 P.M.	27	1 P.M.	26
2 P.M.	27	2 P.M.	25
3 P.M.	28	3 P.M.	23
4 P.M.	27	4 P.M.	24
5 P.M.	26	5 P.M.	24
6 P.M.	26	6 P.M.	23
7 P.M.	26	7 P.M.	23
8 P.M.	25	8 P.M.	23
9 P.M.	24	9 P.M.	23
10 P.M.	23	10 P.M.	22
11 P.M.	23	11 P.M.	21
12 midnight	23	12 midnight	21

CITY OF GREENLAWN HOURLY TEMPERATURE READINGS (Degrees Celsius)			
May 25		**May 26**	
4 A.M.	23	4 A.M.	20
5 A.M.	22	5 A.M.	20
6 A.M.	22	6 A.M.	20
7 A.M.	22	7 A.M.	20
8 A.M.	23	8 A.M.	22
9 A.M.	24	9 A.M.	24
10 A.M.	25	10 A.M.	24
11 A.M.	26	11 A.M.	25
12 noon	26	12 noon	26
1 P.M.	27	1 P.M.	26
2 P.M.	27	2 P.M.	27
3 P.M.	27	3 P.M.	27
4 P.M.	27	4 P.M.	26
5 P.M.	27	5 P.M.	27
6 P.M.	27	6 P.M.	26
7 P.M.	26	7 P.M.	25
8 P.M.	26	8 P.M.	25
9 P.M.	25	9 P.M.	24
10 P.M.	25	10 P.M.	24
11 P.M.	24	11 P.M.	23
12 midnight	24	12 midnight	23

1   2

Take a look at the temperatures in two imaginary cities, Fairhaven and Greenlawn.

The chart shows a record of temperatures for 2 days in Fairhaven. 1

1. On which day did the temperature climb the highest? What was it? At what time?

2. On which day was the temperature lowest? What was the temperature? At what time?

3. On May 26, it rained from 1 P.M. to 2 P.M. What happened to the temperature then?

4. Look at the temperature readings for the city of Greenlawn for the same days. 2 Greenlawn is 200 kilometers (125 miles) east of Fairhaven.

5. Which of the days was cooler in Greenlawn? What was the lowest temperature reached?

6. Compare the temperatures on May 25 in Fairhaven to those on May 26 in Greenlawn. They are very similar. Remember that Greenlawn is east of Fairhaven. With that in mind, predict the weather in Greenlawn for May 27.

✔ Check yourself

1. What are large volumes of air called?

2. What happens when a hot air mass hits a cold air mass?

3. Name the line where two air masses meet.

4. Identify the instrument that measures temperature.

Humidity

On muggy days, there is a lot of moisture in the air. It often makes us uncomfortable. The relative humidity (hyo͞o **mĭd′ĭ** tē) is the amount of water vapor in the air compared with the amount of water vapor the air can hold at that temperature. At 100 percent relative humidity, the air can hold no more. If the air cools and the relative humidity is 100 percent, water vapor begins to change back to a liquid. You sometimes see this as dew. 1 If the temperature is below freezing, frost forms. 2

Warm air can hold more moisture than cold air. That is why we often notice high humidity during warmer weather. In the summer, we usually feel more comfortable after it has showered or rained. 3 This is because the air is drier. Even though it is still hot, we are more comfortable.

Air pressure

Air has weight. It causes a force or pressure. Air presses on everything. We live with air pressure all our lives. Air pressure can be measured.

Air pressure depends a lot on the amount of water in the air. When water is added to the air it makes the pressure less. That's right, <u>less</u>. The drops of water in the air push apart the particles of air. The drops of water weigh less than the particles of air they replace. That means that the more water air contains, the lighter it is. High air pressure means fair weather. When air pressure gets low, it usually means rain.

How can we measure air pressure? There is an instrument that can measure the changes in the weight of the air. It is called a **barometer** (bə rŏm′ĭ tər). 4 In a weather report, you might hear something like this: "The barometric pressure is rising." This means there is less and less water in the air. It usually means the weather will clear.

4

barometer:
An instrument for measuring air pressure.

✓ Check yourself

1. Name three forms of precipitation.

2. When water is added to air, does air pressure become greater or less?

3. What type of weather occurs when air pressure is high?

4. What instrument measures air pressure?

How clouds form

Clouds are masses of water droplets or ice crystals that are suspended in the air. When the droplets or crystals become too heavy to remain suspended, they fall to the Earth.

Clouds form when water vapor in the air condenses. This happens when the air contains more water vapor than it can hold. If air suddenly becomes cold when the relative humidity is near 100 percent, then water vapor usually condenses on particles in the air. They may be dust particles. Or they may be smoke, ice crystals, or water droplets.

Cloud shapes

Just by studying a cloud's shape, a meteorologist can tell a great deal about the cloud. There are three basic kinds of clouds.

Cirrus (sĭr′əs) clouds are thin, feathery clouds. They are high in the sky. They are so high that they are made up of ice particles. They are generally white or whitish in color. 1

cirrus clouds:
Thin, feathery
clouds very high in
the sky.

1

2

3

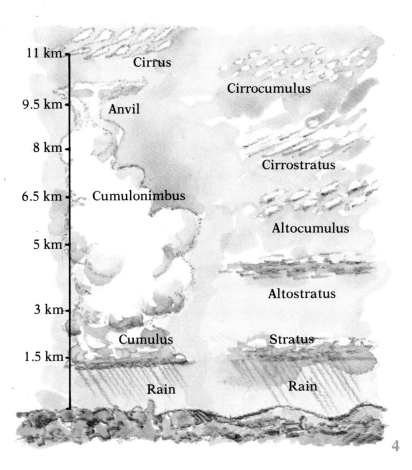

11 km — Cirrus

Cirrocumulus

9.5 km — Anvil

8 km —

Cirrostratus

6.5 km — Cumulonimbus

Altocumulus

5 km —

Altostratus

3 km —

Cumulus   Stratus

1.5 km —

Rain   Rain

4

Cumulus (kyōō′myə ləs) **clouds** look like puffs of cotton piled in a heap. They are commonly known as fair-weather clouds. They are usually nearer the Earth than cirrus clouds. 2

Stratus (strā′təs) **clouds** are low, flat clouds. They are often dark and bring rain. 3

These three cloud types can mix and form different combinations of clouds. The diagram shows the different names and shapes of clouds. 4 It also shows the height at which they are generally found. Read each of their names. It is helpful to know that "alto" means "high." This is why altostratus clouds are higher than stratus clouds.

cumulus clouds:
Big, puffy clouds. They are lower than cirrus clouds.

stratus clouds:
Low, dark, rain clouds.

✓ Check yourself

1. What is a cloud?
2. What clouds are high and feathery?
3. Identify the clouds that look like big cotton puffs.
4. Name the clouds that are low and flat.
5. What are high cumulus clouds called?

Gathering weather data

A place where weather information is recorded is called a weather station. 1 There are hundreds of official weather stations run by the U.S. Weather Service. The station nearest you may be at an airport. Perhaps you can visit it. Then you could see all the instruments.

1

Weather information, or **data**, is collected from stations everywhere. It is fed into a computer. The computer produces a map to show what the weather patterns are. From this, meteorologists predict what the weather will be like.

data:
Information.

Here is a list of weather data. This information comes in from every weather station.

temperature
air pressure
kind and amount of
  precipitation

wind speed and
  direction
relative humidity
cloud cover

When meteorologists record this information on weather maps, they use symbols. For example, a circle that is half filled in means partly cloudy. 2 A system of little flags shows how hard the wind is blowing. 3 Fronts are shown by lines with triangles or half circles on the line. A line with triangles on it shows a cold front. 4 A line with half circles on it shows a warm front. 5 The circles or triangles point in the direction the front is moving. Other types of fronts may also be shown. Other lines show areas that have the same temperature or pressure. 6

2   3   4   5   6

Reading weather maps

Weather maps are made each day. From them, we can follow the movement of air masses. By knowing how air masses move, we can predict what they will do. Look at the map on the next page.

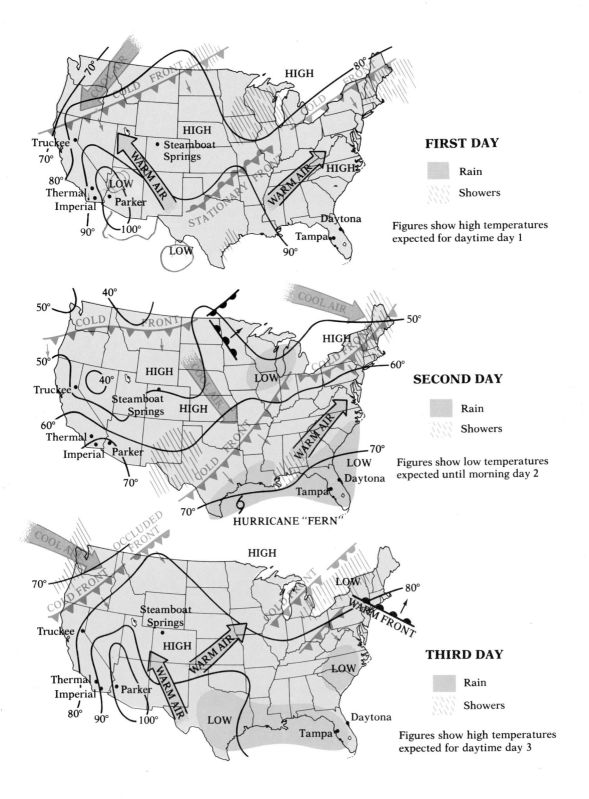

FIRST DAY

Rain

Showers

Figures show high temperatures
expected for daytime day 1

SECOND DAY

Rain

Showers

Figures show low temperatures
expected until morning day 2

HURRICANE "FERN"

THIRD DAY

Rain

Showers

Figures show high temperatures
expected for daytime day 3

Activity

Using weather maps

Use the weather maps to understand how weather systems move. These maps show the weather in the United States for 3 days in September. Look at the first map.

1. How many high pressure systems do you see?

2. How many low pressure systems?

3. How many cold fronts are there?

4. What was the temperature in Parker? In Truckee? Why were they so different?

5. Now look at the second map. The cold front came through Truckee and Steamboat Springs. How can you tell?

6. Find hurricane Fern. Why did parts of Texas and Florida have heavy rain?

7. Look at the third map. What do you think happened to hurricane Fern? Why?

The first map showed three high pressure systems, two lows, and two cold fronts. Because of the cold front, Truckee's temperature was about 85°F (30°C) while it was over 100°F (38°C) in Parker. When the cold front passed through, the temperatures dropped. Meanwhile, hurricane Fern was off the Texas coast causing heavy rain. The next day, pushed by a cold front, it went out to sea.

Now that you can read these weather maps, why not try one from your local paper.

Check yourself

1. Name a place where weather data is recorded.

2. What does a computer produce from this data?

3. How is data recorded on a weather map?

4. What can a weather map tell us about tomorrow's weather?

Science in our world

Weather has always been important to people. It affects the kinds of food that are available. It also affects the kinds of clothing and shelter people need to live comfortably.

The success or failure of many types of work directly depends on the weather. Agriculture, for example, is affected by weather. Farmers depend on the reports of meteorologists. Knowing what the weather is going to be helps the farmers plan carefully. Pilots also must rely on the reports of meteorologists and weather forecasters.

Other people whose jobs are affected by weather are owners of vacation resorts, mail carriers, bus drivers, and baseball players. Now you can understand the importance of accurate weather forecasting.

Careers

Weather forecasters are often **meteorologists**. 1 Meteorologists have to be very careful observers. They have learned to see weather changes and to keep records. In these records, they find patterns. They are good at working with numbers. You might be interested in becoming a meteorologist. Most meteorologists have completed 4 years of college.

1

Review the main ideas

Air travels in masses. The edge of an air mass is called a front. When a warm air mass and a cool air mass meet, there is usually a change in weather conditions. Weather systems generally move from west to east in the United States.

Temperature, relative humidity, and air pressure are important aspects of weather. Precipitation is the general name given to the different forms of moisture that fall from clouds.

Cloud formations help meteorologists predict weather. Clouds of different shapes are found at different altitudes above the Earth. The three basic cloud types are cirrus, cumulus, and stratus.

Chapter review

Science vocabulary

Choose the items in column B that best match the terms in column A.

A

B

1. air mass
2. front
3. barometer
4. cumulus
5. cirrus
6. stratus
7. data

a. puffy clouds piled in heaps
b. information
c. measures air pressure
d. high, feathery clouds
e. the line where two air masses meet
f. low, flat clouds
g. a large volume of air

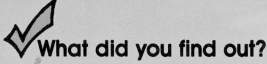

What did you find out?

Answer the following.

1. Air masses, or frontal systems, usually move from _____ in the United States.
   (a) west to east   (b) east to west   (c) north to south

2. Most changes in weather are brought on by
   (a) precipitation   (b) fronts   (c) clouds   (d) data

3. Relative humidity refers to the percentage of _____ that is in the air compared to the amount the air can hold.

4. What happens to the air pressure before it rains?

5. Hail, rain, and snow are all forms of _____ .

6. In order for clouds to form, there must be water vapor and _____ to collect water.

Think and write

1. What does "The relative humidity is 60 percent" mean?

2. What happens when a mass of warm air meets a mass of cold air?

3. Name three kinds of clouds. Describe each, and tell what kind of weather each type usually brings.

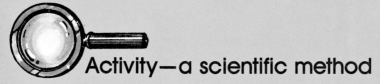

Activity—a scientific method

Purpose: to understand how a cloud forms.

Materials needed

1-gallon clear jug   1-hole rubber stopper to fit jug

15 centimeters (6 inches) of plastic tubing to fit stopper

Procedure

1. Put 10 milliliters of water into a gallon jug. Shake it up.

2. Ask your teacher to drop a smoking match into the jug.

3. Put the tubing into the stopper. Then put the stopper in the jug's mouth. Blow hard through the tubing. Then close it with your finger.

4. Release the tubing and watch the jug. Record what happens.

5. Repeat steps 3 and 4.

Observations

1. Why did you shake the jug?

2. What happened when you released the tube?

3. Why do you think the smoking match was put in?

4. What three things do we need for a cloud to form?

14

Our solar system

In this chapter, you will find out that

■ The sun is necessary to life on Earth.

■ Earth's one satellite, the moon, appears to change shape.

■ Earth is one of nine planets, some of which also have moons.

■ Our solar system includes a sun, nine planets, their moons, and other heavenly bodies.

Before you begin

The science of **astronomy** (ə **strŏn'ə** mē) has interested people throughout history. Astronomy is the study of all the heavenly bodies, including the sun, the moon, the planets, and the stars. Do you want to learn about the solar system we live in?

When was the last time you gazed up at the night sky and saw the moon? Do you remember its shape? Was it big and round, or could you see only a tiny slice of it? Perhaps there have been other times when you could not find the moon at all. Why does the moon change its shape?

What is the difference between a planet and a moon? Did you know that while the Earth has only one moon, other planets in the solar system have as many as seventeen? Did you know that the sun is a star? Compared to other stars, the sun is not even a very bright star. Why, then, does it seem so bright to us?

astronomy:
The study of heavenly bodies.

213

A very special star

Have you ever seen a sun at night? You have if you have seen the stars twinkling in a clear sky. These stars are really suns. Many are like our sun. They are all balls of glowing gases. Some stars are much larger than our sun. Others are smaller. What makes our sun such a special star? 1

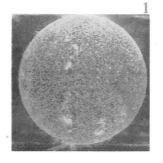

For one thing, there would be no life on Earth without the sun. The glowing gases that make up the sun are always giving off energy. This energy comes to us as sunshine that brightens our day. With this light, plants are able to grow and produce food. All living things depend on the sun. 2

The sun's energy also warms the air, the land, and the water. The warm air causes winds to blow. The sun helps make our weather.

The sun is important for another reason, too. It keeps Earth from going off into space. That does not mean that Earth is standing still. Our planet is moving all the time. It spins, or **rotates**, on its axis. But while Earth is rotating, it is also moving in a path, or **orbit** (ôr′bĭt), around the sun. 3   Earth stays in its orbit because the sun holds it by **gravity** (grăv′ĭ tē). Gravity is a force that pulls things toward each other. Earth's gravity helps keep you on the ground. The sun's gravity keeps planet Earth in its orbit.

rotate:
To move around a center, or axis.

orbit:
The path an object takes around another object.

gravity:
A force that tends to pull things toward each other.

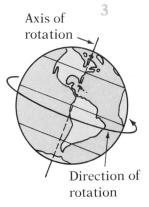

Axis of rotation

3

Direction of rotation

Check yourself

1. What are stars?
2. What is Earth's path around the sun called?
3. How does the sun help make our weather?
4. Name the force that tends to pull things toward each other.

Our nearest neighbor

When the moon is full and the night is clear, you feel, sometimes, that you can reach out to the moon and almost touch it. It may be impossible to touch, but the moon is our closest neighbor in space. It is about 384,000 kilometers (240,000 miles) from Earth. 1 The moon is not a ball of gases like the sun. As the astronauts have seen, the moon is a rocky and barren place. 2

Just as Earth revolves in an orbit around the sun, the moon revolves in an orbit around Earth. The

moon completes an orbit around Earth about once every month. The moon has no light of its own. It appears to have light. But moonlight is just the reflection of the sun's light on the moon's surface. As the moon moves around Earth, we are not able to see all the surface that the sun lights up. This makes the moon appear to change shape. These changing shapes are called its **phases**.

When the moon is in position 1 in the diagram, we cannot see it. 3 The sun's light is reflected on the side of the moon that is facing away from Earth. We call this phase the new moon. As the moon travels to position 2, we see a crescent moon.

phases:
The different shapes the moon appears to have at different times.

3

Phases of the Moon

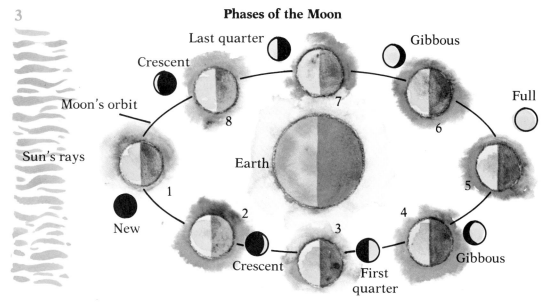

Key: The sun's rays light one side of the moon. The yellow shows what you see from Earth.

The moon continues to travel until we can see all of its lighted surface. We call this a full moon. The moon continues to orbit around Earth, and, as it does, we see less and less of it. Finally, it is back to position 1, and the cycle begins again.

Activity

Phases of the moon

1. See if you can learn all the phases of the moon. Two of them are easy: the new moon, which cannot be seen, and the bright, full moon. There are two times when the moon has a crescent shape. Whenever the moon is halfway between new and full, we call it a quarter moon. If it is on its way to becoming full, it is a first-quarter moon. If the moon is getting smaller, it is a last-quarter moon. Just before and after a full moon, it is in the gibbous phase. "Gibbous" means "hump."

2. The diagram shows the phases of the moon for one month. **1** Use the names from the list at right to identify each phase below.

new crescent
old crescent
new moon
full moon
new gibbous
old gibbous
first quarter
last quarter

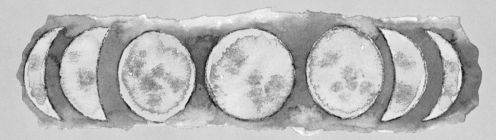

1

The phases of the moon are caused by its orbit around Earth. Any object that orbits a larger object is called a **satellite**. You may have heard of satellites before. There are many satellites orbiting Earth. The moon is our only natural satellite. Earth is in orbit around the sun. Is Earth a satellite of the sun?

satellite:
An object that orbits, or revolves around, another object.

Check yourself

1. How long does the moon take to orbit Earth?

2. What causes the phases of the moon?

3. During what phase of the moon can we see all of its lighted surface?

4. Define "satellite."

The planets

Have you ever seen a photograph of Earth taken from far above? 2 Earth is a **planet**. A planet is a body in space that does not give off light. A planet gets its light from the star around which the planet revolves.

planet:
A body in space that orbits a star or sun. It has no light of its own.

2

Earth is a planet that revolves around the star we call the sun. There are eight other planets that also revolve around the sun. The major planets are named Mercury, Venus, Earth, Mars, Jupiter, Saturn, Uranus, Neptune, and Pluto. The chart lists each of the planets and gives you an idea of their sizes.1   The planets are shown in their order from the sun. Notice that Mercury is the closest planet to the sun.

1

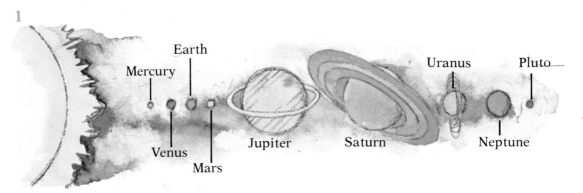

Some of the planets are bright enough to be seen without a telescope. They are Venus, Jupiter, Saturn, and Mars. Mars can also be recognized by its reddish color. 2   The light planets reflect from the sun does not twinkle the way star light seems to do. It is a steady light.

2

3

Some of the planets have another interesting feature. Jupiter, Saturn, and Uranus have rings around them. The photograph is of Saturn and its rings. 3 The rings are made up of frozen gases and particles. Recent explorations in space have given scientists a better idea of how large these rings are and what they contain.

✔ Check yourself

1. What is Earth?
2. How many planets revolve around the sun?
3. How are Jupiter, Saturn, and Uranus alike?

Other planets and their moons

Earth is not the only planet with a moon. In fact, some of the other planets have more than one moon. The planets Mercury and Venus have no moons. Pluto, like Earth, has one. Mars has two moons. Neptune has five. Uranus may have as many as five. We will learn more about Uranus when Voyager 2 reaches the seventh planet from the sun in 1986.

The larger planets seem to have the most moons. Jupiter, the largest planet in our solar system, has 16 known moons. Saturn, the second-largest planet, has 17 known moons. One of these, Titan, is the largest of Saturn's moons. 1 Recent satellite explorations have turned up new facts about the moons of other planets.

1

Activity

A trip to the planets

Imagine you are about to take off in the space shuttle. **2** But first, make sure you have an idea of what is out there. Look back to the chart and see if you can answer all these questions.

1. Which planet is closest to the sun?
2. Which planet is farthest from the sun?
3. Between which two planets do we live?
4. Why do you think that Earth and Venus are sometimes called "twin" planets?
5. Which two planets are smallest?
6. Which planet has a diameter about one-tenth that of the sun?
7. Now you are to visit each of the following planets. Identify each planet by its description.
   (a) large, has rings, 17 moons
   (b) small, no moons, closest to sun
   (c) close to Earth, reddish color
   (d) far from sun, has five moons
   (e) has two moons
   (f) very small, very far away
   (g) only planet to have life as we know it
   (h) can be seen without telescope, no moons
   (i) has 16 moons, very large

Now, use this chapter to check your answers. How did you do? Are you getting familiar with our neighbors, or did you get lost in space?

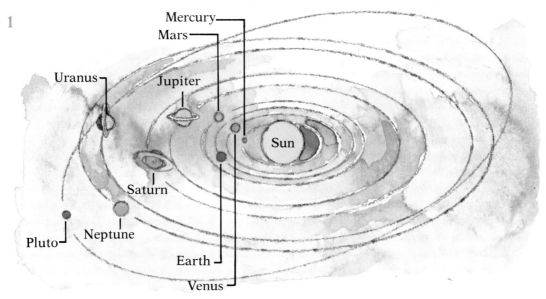

1 Mercury
Mars
Uranus
Jupiter
Sun
Saturn
Pluto
Neptune
Earth
Venus

The solar system

A system (**sĭs′təm**) that includes one or more suns and planets and their moons is called a **solar system**. "Solar" means "related to the sun." So far, no one has found another solar system quite like ours. But scientists believe that they do exist.

Our solar system is made up mostly of space. Within that space are the sun, the nine planets, and their moons. The sun's force of gravity holds the nine planets and their moons in orbit. 1

Far from Earth, between the orbits of Mars and Jupiter, there are many tiny bodies. These miniplanets are known as **asteroids (ăs′tə roidz′)**. Some asteroids can be seen with a telescope. There are more than 1600 of them. Some are as small as 2 kilometers (1.2 miles) in diameter. Others resemble flying mountains.

Once in a while, bodies called **comets (kŏm′ĭts)** travel through the solar system. 2 A comet is thought to be made up of a combination of gases, dust, and ice. A comet is recognized by its long tail.

solar system:
A system that includes one or more suns, planets and their moons, and other bodies.

asteroid:
Tiny bits of planet-like material that orbit the sun; a miniplanet.

comet:
A body thought to be made up of gases, dust, and ice and having a "tail."

224

2

Some comets have returned to our solar system so regularly that predictions can be made about when they will return. Halley's comet is a large comet that returns every 76 years. The next year Halley's comet will return is ~~1986~~.

Meteoroids (mē'tē ə roidz') are stony, solid particles that travel through space. When a meteoroid comes close to Earth, it gets hot and glows. It leaves a trail of light as it crosses the sky. This trail of light is called a meteor (mē'tē ər). Often meteors are called falling stars or shooting stars, even though they are not really stars. The meteoroid usually burns up before it hits the ground. If it falls to Earth, its rocklike pieces are called meteorites (mē'tē ə rīts').

meteoroid:
A stony particle that travels through space.

✓ Check yourself

1. Which planets have moons?
2. Define "solar system."
3. What is another name for the miniplanets?
4. Describe the size of miniplanets.
5. List the materials that are thought to make up a comet.
6. What is the difference between a meteoroid and a meteorite?

Science in our world

Through space exploration, many important discoveries have been made. Samples of moon rocks have given scientists clues as to how Earth and its moon came into existence.

Space vehicles have landed on Mars. Others have traveled around Venus. Some have gone on trips past the planets Jupiter and Saturn. Thanks to computers and cameras, close-up photographs of the planets have been transmitted to Earth. Other space vehicles called satellites stay closer to Earth. Some take weather pictures. Others help send TV signals across the ocean.

Careers

Astronomers (ə **strŏn**'ə mərz) are scientists who study objects in space. These objects include the sun, the moon, the planets, stars, meteoroids, asteroids, and all other heavenly bodies.

Astronomers use telescopes to observe objects in space. Today many astronomers also use devices such as radio telescopes and computers. They also get information from satellites and space probes.

Carl Sagan is a famous astronomer. 1 Perhaps you have seen him on TV. Sagan uses information sent to Earth from space probes to understand the planets. He has discovered many things about the atmospheres of some of the planets. This information may be very useful for exploring space in the future. Sagan is also searching for signs of life in other parts of the universe.

1

Review the main ideas

The sun is a star of special importance. It is the center of our planetary system. It gives off the light and heat that make life possible. Earth's moon is our nearest neighbor in space. It does not give off light but reflects light from the sun. It appears to change its shape as it orbits Earth.

Earth is one of nine planets that circle around the sun. Seven of the planets have moons. Nine planets, their moons, the sun, and the space surrounding them make up our solar system. Also within the solar system are asteroids, comets, and meteoroids.

Chapter review

Science vocabulary

Choose the items in column B that best match the terms in column A.

A	B
1. astronomy	a. one or more suns, planets and their moons, and other bodies
2. orbit	
3. rotate	b. stony particles that travel through space
4. satellite	c. an object that revolves around another object
5. planet	d. spin
6. solar system	e. miniplanets
7. asteroids	f. study of objects in the sky
8. comets	g. formed of gas, dust and ice; have tails
9. meteoroids	h. path of a satellite
	i. a body in space that does not give off light

✓ What did you find out?

Answer the following.

1. Scientists who study objects in space are called _____ .

2. To us on Earth, the most important star is the star called the _____ .

3. Earth rotates on its axis and _____ around the sun.

4. Our nearest neighbor in space is the _____ .

5. A full moon is an example of one of the moon's _____ .

6. Including Earth, the number of planets in our solar system is _____ .

Think and write

1. Name two reasons why the sun is important.
2. Why do we not see a full moon every night?

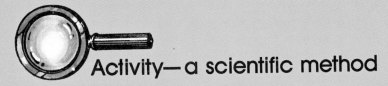

Activity—a scientific method

Purpose: to observe the stars.

Materials needed

an evening   a clear sky   a telescope (optional)

Procedure

1. Go out after sunset. Do not wait for complete darkness.
2. The first object you see at twilight is called the evening star. It is actually a planet, either Venus or Mars.
3. Find out which planet is the evening star that month. Check your newspaper or call the Weather Bureau.
4. If you have a telescope, use it. Notice the color of the evening star.
5. Now look for real stars. Compare the way they look to the way the evening star looks. Record your observations.
6. Go back out in an hour. Try to find the evening star again. Count how many stars you see. Notice their colors.

Observations

1. What color was the evening star? What planet was it?
2. Which was brighter: the evening star or real stars? Explain.
3. How many stars did you count the first time? The second time?

15

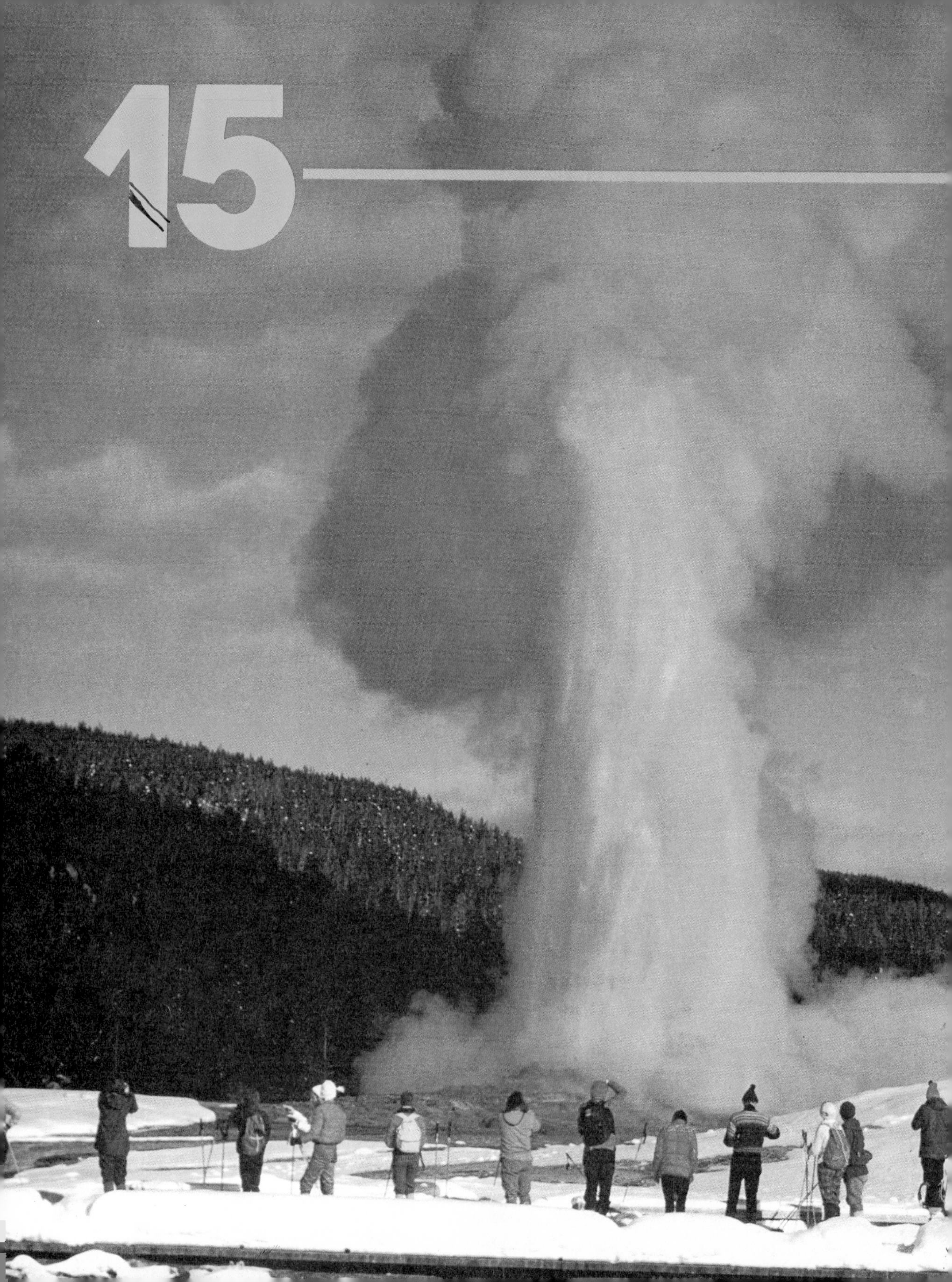

New sources of energy

In this chapter, you will find out that

- People use heat from inside the Earth.
- We can get electricity from garbage and wind.
- The sun's energy can be changed to heat and electricity.
- Nuclear reactors use energy from atoms.

Before you begin

The world's supply of fossil fuels is running out. Oil is not as plentiful as it once was. People have burned it much faster than it can be replaced. People are now looking for new sources of energy.

Wind, garbage, and the atom are some of these new energy sources. People are also using heat trapped inside the Earth. Rocks deep in the Earth are very hot. In some places, groundwater comes in contact with these rocks. The water is heated and flows to the surface, forming a hot spring. Did you know that a geyser is a kind of hot spring that sometimes erupts? It sends up columns of hot water and steam into the air. Do you know how a geyser may be used to produce electricity?

Using geothermal energy

In most parts of the world, **geothermal** (jē′ō-thûr′məl) heat sources are not near the surface of the Earth. They are too deep to reach. But in some places, the heat is near the surface. It can be used for electricity and heating.

People have built geothermal power plants at The Geysers in northern California. 1 These plants collect steam from **geysers**. The steam is used to turn a turbine in an electric generator.

geothermal:
The kind of heat that comes from within the Earth.

geyser:
A hot spring that sometimes erupts.

1

In Iceland, there is a lot of geothermal hot water. The Icelanders use this water to heat their homes and even their swimming pools. More than 90 percent of the homes in the capital city, Reykjavik, are heated by geothermal hot water. The water is piped from wells to the homes. Heating homes with geothermal hot water is also done in some places in the western United States. Klamath Falls, Oregon, has heated homes from hot water wells for many years.

Garbage power

Many tons of trash are produced in the United States every day. This trash ends up in a garbage dump. Our garbage dumps are filling up. Where will we put our garbage in the future? 2

Most of the trash we produce could be burned. A growing number of communities are experimenting with this idea. The heat from the burning trash turns water into steam. The steam is used to turn a turbine in an electric generator.

2

Check yourself

1. What is heat energy deep inside the Earth called?

2. What can geothermal hot water be used for?

3. Where in the U.S. is geothermal hot water used to heat homes?

Energy from the sun

The sun's energy is called **solar energy (sō′lər ĕn′ər jē)**. It can be used to heat buildings. This building has **solar collectors (sō′lər kə lĕk′tərz)** on its roof. 1 These collectors absorb the solar energy and change it into heat energy. Air or water passes through the collectors. The air or water absorbs the heat and carries it to the rest of the building.

solar energy:
Energy from the sun. It reaches Earth as light and heat.

solar collectors:
Panels used to catch solar energy.

1

Activity

Comparing temperatures

1. Get two plastic cups, two thermometers, some ice water, and a watch or clock.

2. Pour equal amounts of ice water into the cups.

3. Place a thermometer in each cup.

4. Put one cup in the sun. Put the other cup in the shade.

5. Read the temperature on each thermometer each minute for 15 minutes. Record the temperatures.

6. Make a graph. The horizontal axis represents time in minutes. The vertical axis represents degrees of temperature. Place the data from your activity on the graph.

7. What does the graph show?

Solar energy can be changed to electric energy. Collectors absorb the sun's energy and become hot. This heat is transferred to water. The water becomes hotter and hotter. Eventually, it turns to steam. The steam can be used to turn a turbine in an electric generator.

It also is possible to change solar energy directly into electricity. This is done by a device called a **photovoltaic cell** (fō′tō vŏl tā′ĭk). Each tiny cell produces a very small amount of electricity. But by putting a group of cells together, much more elec-

photovoltaic cell:
A device that changes solar energy directly into electric energy.

tricity can be produced. A group of photovoltaic cells make up a **solar battery (sō′lər băt′ə rē).** The electricity produced is used to operate satellites. **1** Solar batteries also are being used on Earth.

solar battery: A group of photovoltaic cells that are connected together.

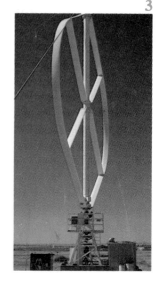

Using wind energy

For centuries, people have put the wind's energy to work. They have used the wind to move sailing ships across the sea. They have used the wind to operate windmills. The windmills, in turn, ground corn, lifted water out of the ground, and did other jobs. **2**

Now, people are experimenting with a new kind of windmill called a wind turbine, or wind generator. There are many different designs. **3** These generators change the wind's energy into electricity. They may become important sources of electricity in places that have strong, steady winds. Do you think these generators would be practical where you live?

✓ Check yourself

1. What is the sun's energy called?
2. Name a device that changes solar energy into heat energy.
3. What type of energy do solar batteries produce?
4. How has wind energy been used in the past?
5. What is being used now to change wind energy into electricity?

Energy from atoms

As you know, there are other sources of energy besides the sun. Scientists are always looking for different sources to help supply energy needs.

All matter is made up of tiny particles called **atoms (ăt′əmz).** 4 The center of an atom is called the **nucleus (nōo′klē əs).** If the nucleus is split, energy is released. This energy is called **nuclear energy (nōo′klē ər ĕn′ər jē).**

Uranium is a heavy metal. It is found in the Earth's crust. During the Second World War, scientists discovered how to get energy from uranium. They found out that splitting uranium nuclei releases a lot of heat energy, or nuclear energy. Splitting the nuclei in 1 kilogram (2.2 pounds) of uranium produces as much heat energy as burning 3 million kilograms (6,600,000 pounds) of coal.

atoms: Tiny particles of which all matter is made.

nucleus: The central part of an atom.

nuclear energy: Energy released when atoms are split.

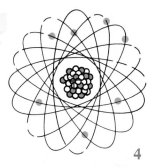

4

● = Neutron

o = Proton

● = Electron

In a nuclear power plant, the uranium is contained in special **fuel rods (fyōō′əl rŏdz).** 1 These rods are placed in a device called the **nuclear reactor (nōō′klē ər rē ăk′tər).** As the nuclei split, the heat they release is used to heat water. The water turns to steam. The steam is used to turn a turbine in an electric generator. 2

fuel rods:
The part of a nuclear power plant that contains the uranium.
nuclear reactor:
The device in which nuclei are split to get nuclear energy.

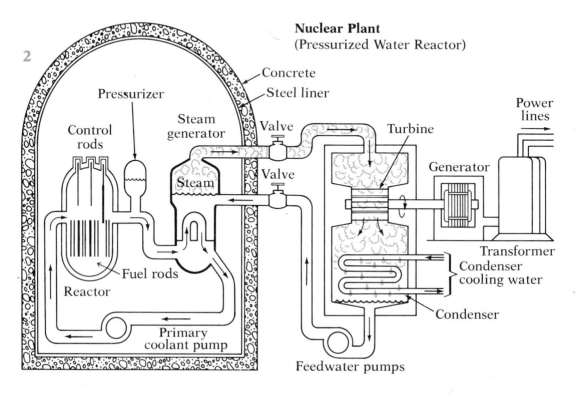

2

Nuclear Plant
(Pressurized Water Reactor)

Concrete
Steel liner
Pressurizer
Control rods
Steam generator
Valve
Steam
Valve
Turbine
Generator
Power lines
Fuel rods
Reactor
Primary coolant pump
Feedwater pumps
Condenser
Condenser cooling water
Transformer

Activity

Reading about nuclear energy

1. Find a news article or an editorial on nuclear energy.
2. What is the writer's opinion of nuclear energy? Can you tell? How?
3. Write a report based on your findings. Illustrate it with art or photographs.
4. Draw a cartoon based on the article.

Radiation

The releasing, or giving off, of energy is called **radiation** (rā′dē ā′shən). Heat is one kind of radiation released when uranium atoms are split.

But other kinds of radiation are also released. These kinds of radiation are in the form of particles and rays. Some of these can get inside organisms and cause serious damage to cells. This kind of radiation can be very dangerous. It can kill living things and cause diseases.

Some forms of radiation are helpful. X-rays are used to find broken bones. They also help diagnose diseases such as tuberculosis. Some hospitals give patients radiation treatments. 3 Radiation treatments can kill certain growths or tumors. Before radiation treatments, most tumors had to be removed by surgery. Radiation can reach some places where a scalpel cannot.

radiation:
The releasing, or giving off, of energy.

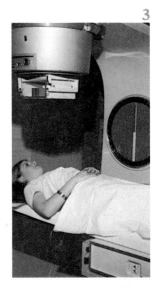
3

Check yourself

1. Name the type of energy released when atoms are split.
2. Identify the metal used to release nuclear energy in power plants.
3. Can radiation harm living things?
4. How do some forms of radiation help us?

Science in our world

Using the energy stored in atoms is very controversial. Some people want to use it. Other people are against using it.

People who support nuclear power plants think nuclear energy is the best answer to our energy needs. They say it is safe. They say there is very little danger that radiation will get into the environment. They say nuclear power causes less pollution than other sources of energy.

Other people do not think nuclear power plants are safe. They also point out that no safe way has been designed to get rid of the wastes from nuclear power plants. These wastes are dangerous. They give off harmful radiation.

Special shells are built around nuclear reactors to prevent radiation from escaping into the environment. People who work in nuclear power plants wear protective clothing.

Careers

Nuclear power plants produce energy by controlling nuclear reactions. Only controlled nuclear energy is useful for generating power. Most of what we know about nuclear power comes from the work of **physicists** (fĭz' ĭ sĭsts). Physicists are scientists who study matter and energy.

Enrico Fermi was a famous physicist. 1 He built the first nuclear reactor. Fermi taught us how nuclear power could be controlled. Today, nuclear reactors supply part of our energy. Radiation produced during nuclear reactions can be dangerous if it is not handled properly. Scientists and engineers are working to find ways of making nuclear power safer to use.

1

Review the main ideas

People are looking for ways to replace fossil fuels. They are testing a variety of energy sources. The heat deep inside the Earth can be used to heat buildings. Or it can be turned into electricity. Electricity also can be produced from burning trash, solar energy, wind power, and nuclear energy. Nuclear energy is the most controversial of these new energy sources. Splitting the nucleus of an atom releases harmful radiation.

Chapter review

Science vocabulary

Choose the items in column B that best match the terms in column A.

A	B
1. geothermal energy	a. changes sunlight directly into electric energy
2. geyser	
3. solar energy	b. absorbs sun's heat
4. solar collector	c. contains uranium
5. photovoltaic cell	d. heat in the Earth
6. solar battery	e. made up of many cells
7. atoms	f. atom's center
8. nucleus	g. sun's energy
9. fuel rod	h. uses fuel rods
10. nuclear reactor	i. contained in all matter
	j. hot spring that erupts

✓ What did you find out?

Answer the following.

1. At The Geysers in California, _____ energy is being changed into _____ energy.

2. Solar batteries change solar energy directly into _____ .

3. Wind energy is changed into electric energy by _____ .

4. The energy released by splitting the nucleus of an atom is called _____ energy.

5. In nuclear reactors _____ are split to produce energy.

6. The giving off of energy is called _____ .

Think and write

1. Explain how garbage can be used to produce electricity.

2. Where would wind turbines work best?

3. Why are special shells built around nuclear reactors?

4. Write a report on radiation. Include its dangers and benefits.

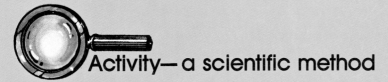

Activity—a scientific method

Purpose: to show how the sun's heat energy affects a thermometer.

Materials needed

4 mirrors   1 thermometer   modeling clay
black construction paper   cellophane tape   a sunny day

Procedure

1. Tape a piece of black paper over the entire thermometer bulb.

2. Put clay on the bottom of the thermometer so that it stands up. Use clay to make a base for standing up each mirror.

3. Stand the thermometer on the window ledge. Its back should face the sun. After 5 minutes, record the temperature.

4. Stand a mirror in front of the thermometer to reflect sunlight onto the paper. After 5 minutes, record the temperature.

5. Add another mirror and repeat step 4. Repeat steps 4 and 5 twice more.

Observations

1. What happens each time you add another mirror? Where is the energy coming from to heat the thermometer?

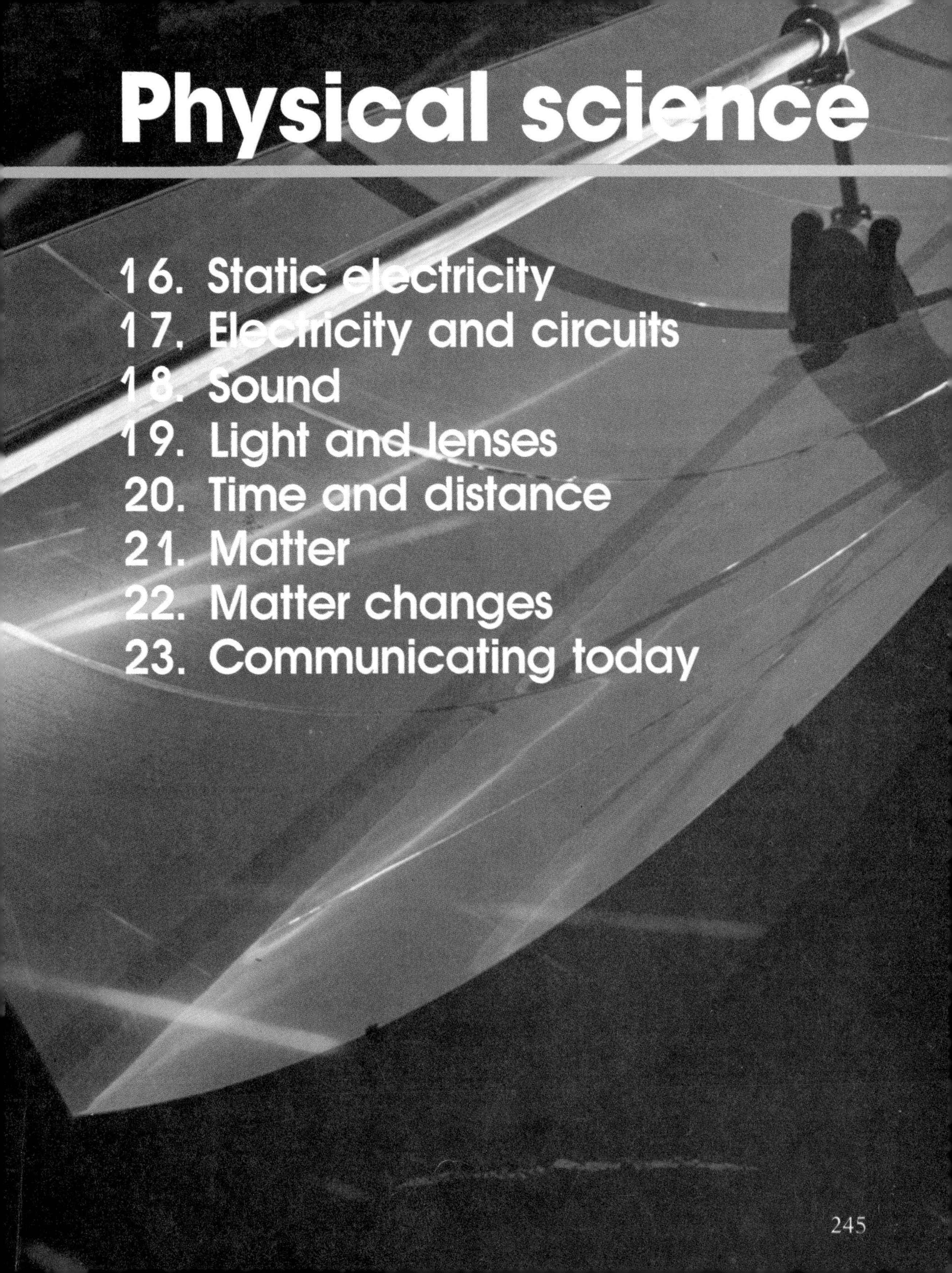

Physical science

16. Static electricity
17. Electricity and circuits
18. Sound
19. Light and lenses
20. Time and distance
21. Matter
22. Matter changes
23. Communicating today

245

16

Static electricity

In this chapter, you will find out that

- There is a form of electricity called static electricity.
- Objects can be charged with electricity.
- Like charges repel each other and unlike charges attract each other.
- Objects can lose their electric charge.

Before you begin

You have probably seen lightning streaking across the sky. Maybe you have felt a shock after walking across a rug and touching a metal object. Perhaps you have been annoyed at the way your hair stands on end after you have washed it, especially on a dry day. All of these things are the result of static electricity.

What is static electricity? What causes it? How does it work? Is this kind of electricity useful, or is it just a nuisance?

Did you know that your body can be charged with static electricity but that a wire cannot be? That is just one of the odd things about static electricity.

247

Electricity at rest

The word "static" means "not moving." **Static electricity** is electricity at rest. When an object contains static electricity, we say that the object is charged. What does this mean?

All matter contains tiny electrical bits, or particles. The electrical property of these particles is called a **charge**. Sometimes, an electrical charge builds up on a certain material that is a poor conductor. The charge is static, or at rest. The charge remains that way until it is caused to move. It is possible to produce an electric charge on certain materials.

static electricity:
Electricity at rest.

charge:
The electrical property of particles of matter.

Activity

Making an electric charge

1. You will need a sheet of paper and a pencil.
2. Put the paper on your desk or on a tabletop. Now rub the pencil across the paper. Do this eight to ten times.
3. Slowly pick up a corner of the paper. 1 What happens? Is the paper electrically charged? Does it contain static electricity?
4. Repeat steps 2 and 3. Rub the paper with a piece of wool, a ruler, a glass, a coin, a crayon box, and anything made of metal.
5. Make a chart, listing the objects you tested and whether or not they produced an electric charge. 2

Object Tested	Electric Charge Produced	No Electric Charge Produced
Pencil		
Wool cloth		
Wooden ruler		
A glass		
A quarter		
Crayon box		
Metal ruler		

Whether or not an object can be charged with static electricity depends on the kind of material from which the object is made. For example, most nonmetals, such as glass and wool cloth, are materials that can be charged. Most metals cannot be charged.

Do you think metals are good conductors of electricity? Explain your answer.

Check yourself

1. What name is given to electricity at rest?
2. What do we call the electrical property of particles of matter?
3. Can paper be charged?
4. What kind of material can usually be charged?

Two kinds of charges

When an object is charged, it contains static electricity. The paper Dan used in his experiment did not contain static electricity at first. But after Dan rubbed the pencil over it, the paper was charged. It then contained static electricity.

There are two kinds of electric charges: positive (**pŏz′ĭ tĭv**) and negative (**nĕg′ə tĭv**). Usually, the number of positive electric charges in an object equals the number of negative electric charges. 1 But sometimes an object loses some of its negative charges. Then it has more positive charges, and we say the object is **positively charged**.

If more negative electric charges are collected, then the object is **negatively charged**. When an object is charged positively or negatively, it contains static electricity.

How objects are charged

There are a number of ways an object can be charged so that it contains static electricity. Try this. Rub your hands together briskly for a few seconds. What happens? Do your hands feel warm? The rubbing of one object against another is called **friction** (**frĭk′shən**). Friction <u>always</u> produces heat. And friction <u>sometimes</u> produces static electricity.

Greek scientists discovered what we call friction thousands of years ago. The early Greeks made buttons for their clothes from a yellow-brown substance that we call amber (**ăm′bər**). 2 The Greeks saw that the amber buttons attracted lint, hairs, and other materials. This always seemed to happen after the amber buttons rubbed against wool clothing. This power of attraction was called "the power of amber."

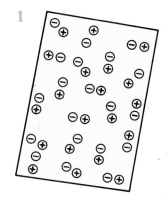

1

positively charged: Having more positive charges than negative charges.

negatively charged: Having more negative charges than positive charges.

friction: The rubbing of one object against another, producing heat and sometimes static electricity.

250

The rubbing of the wool against the amber buttons caused some negative electric charges to rub off the wool onto the amber. The amber buttons became charged with static electricity.

What kind of charge did the buttons have? If you said a negative charge, you are right. What caused the buttons to become charged with static electricity? If your answer is friction, or the rubbing of the wool against the amber, you are right again.

✓ Check yourself

1. Name the two kinds of electric charges.
2. How does an object change when it gains negative charges?
3. What does friction sometimes produce?
4. What happens to amber buttons rubbed against wool?

Friction and static electricity.

You can see what friction sometimes does. Try doing the following activity.

Activity

Producing static electricity by friction

1. You need a rubber comb and some paper.
2. Tear the paper into small pieces. Hold the comb near the paper. What happens? Why?
3. Comb your hair five or more times. Hold the comb near the paper. What happens? Why?
4. Run your hands over the comb and hold it near the paper again. What happens? Why?

At first, the comb had an equal number of positive and negative charges. That is, the comb was not charged. It contained no static electricity. The rubber comb was electrically **neutral (nōō′trəl)**. Because it was neutral, nothing happened when it was brought near the pieces of paper. 1

An electric charge was produced by friction—the rubbing of the hair against the comb. 2 Negative charges in the hair were rubbed off onto the comb. Now the comb contained static electricity. It was negatively charged. The hair was positively charged. When the comb was brought near the paper again, what happened? 3 Why?

neutral:
Having an equal number of positive and negative electric charges.

Running your hands over the comb caused it to lose its electric charge. Once again, the comb was neutral, neither positively charged nor negatively charged. And the pieces of paper were no longer attracted to it.

Later in this chapter, you will learn how objects lose electric charges.

Attract or repel

As we said, there are positive electric charges and negative electric charges. When an object with a negative charge and an object with a positive charge are brought close to each other, they **attract** (ə trăkt′) each other. That is, they pull toward each other. Sometimes, they actually come together. It is as if the negative charges are trying to join up with the positive charges, and vice versa.

When two objects with the same kind of electric charge are brought near each other, they **repel** (rĭ pĕl′) each other. In other words, they try to push away from each other. Like charges repel each other, and unlike charges attract each other. This can be proven in a number of ways.

For example, two glass rods are rubbed with the same piece of wool cloth. One rod is hung from string, and the other rod is held. When the rod that is held is moved close to the hanging rod, the hanging rod moves away from the held rod. 1 Because both rods have the same kind of charge, they repel each other.

When a rubber rod is used in place of one of the glass rods, however, the two rods attract each other. 2 What does this tell you about the charges in the two rods?

attract:
To pull toward each other.

repel:
To push away from each other.

1

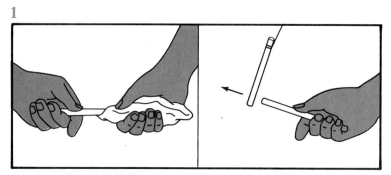

2

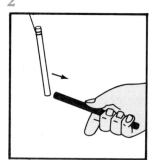

Loss of charges

Charged objects can also lose their electric charges. That is, they can be **discharged** (dĭs **chärjd′**). Some charges are lost slowly. Others are lost very quickly. A balloon charged with static electricity will stick to a wall longer in dry weather than in humid weather. 3   This is because electric charges are carried away more quickly when moisture is present. When an object loses its electric charge, it is neutral again.

When you shuffle across a rug and touch a doorknob, the charge in your body is lost very quickly. That is why you feel a shock. Such a quick discharge of static electricity can cause sparks. You can prove this by doing the rug-and-doorknob test in a darkened room. Watch carefully as you touch the doorknob, and you may see the spark. 4

Lightning is another example of a sudden loss, or discharge, of static electricity. Of course, the spark made by lightning is much more powerful than the spark made when you touched the doorknob.

discharge:
To lose an electric charge.

3

4

✔ Check yourself

1. What is an object with no electric charge called?

2. How do like charges act when they come near each other?

3. How do unlike charges act when they come near each other?

 # Science in our world

Machines that separate charges are used to make copies of documents. They are known as electrostatic copiers. The machines make charged places on a piece of paper. A powder is attracted to the charged places on the paper. Then the paper is heated, and the powder melts and sticks to the paper at the charged places. The charged paper is the copy. Notice the difference between the original photo of the dog and the copy of the photo. 1

Careers

Electricians wire houses so that the people living there will have electricity. 2 They also repair problems with electricity in circuits, but not problems with static electricity. Do the lights work in your home? Does your TV work when you turn it on? Then an electrician has been there.

Electricians are high school graduates. Sometimes they go to trade school. They always start out by helping experienced electricians. That is how they get on-the-job training.

2

Review the main ideas

Static electricity is electricity at rest. Objects that contain static electricity are said to be charged. Certain materials, such as nonmetals, can be charged with static electricity. One way this can be done is by friction, or the rubbing of one object against another.

There are two kinds of charges: positive and negative. Materials that lose negative charges are said to be positively charged. Materials that gain negative charges are negatively charged. An object that contains an equal number of positive and negative charges is electrically neutral.

Objects with like charges repel, or push away from, each other. But objects with unlike charges attract each other.

When objects are discharged, they no longer contain static electricity. They have lost their electrical charges. This discharge can happen slowly or quickly. Moisture in the air can cause a discharge to happen quickly. A very quick discharge of static electricity can be dangerous.

Chapter review

Science vocabulary

Choose the items in column B that best match the terms in column A.

A

1. static electricity
2. charge
3. friction
4. electrically neutral
5. attract
6. repel
7. discharge

B

a. having an equal number of positive and negative charges

b. to pull toward

c. to lose electricity

d. electrical property of particles of matter

e. to push away

f. electricity at rest

g. rubbing of one object against another

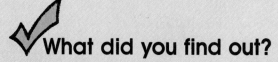

What did you find out?

Answer the following.

1. The result of friction is <u>always</u> _____ .
2. Objects that repel each other have the same _____ .
3. Lightning is an example of electrical _____ .
4. Materials that can be electrically charged are usually _____ .

Think and write

1. If all electricity stopped, how would your life be different?
2. What does current electricity mean? (Clue: Look up the word "current.")

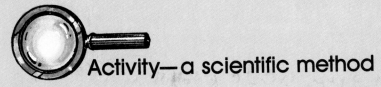

Activity—a scientific method

Purpose: to make a charge detector to test objects for electric charges.

Materials needed

blunt scissors   plastic ruler (with center hole)   straight pin
aluminum foil   soda straw (to fit through ruler hole)
plastic wrap   objects, such as paper, pencil, metal ruler, etc.

Procedure

1. Cut a strip of aluminum foil about 4 centimeters by 20 centimeters.

2. Cut the soda straw in half. Wrap the foil strip around a half-straw.

3. Push a pin through the wrapped straw a few millimeters above its center. Put the straw through the hole in the ruler.

4. Rest the ruler on a few books or a table edge. The straw part should not touch anything.

5. Rub a piece of plastic wrap on your desk. Then touch the straw with it.

6. Rub a new piece of plastic wrap on your desk. Bring it near the charge detector.

Observations

1. What happened when you brought the plastic wrap near the charge detector? What does this tell you about the charge?

2. Repeat the experiment with different objects. Do all the charged objects attract the detector? Do any repel it?

3. Do you get different results at different ends of the detector?

17

Electricity and circuits

In this chapter, you will learn that

- Electricity is a form of energy.
- Other forms of energy can be changed into electricity.
- Electricity flows through circuits.
- Electric circuits may be series or parallel.
- Electricity and magnetism are related.

Before you begin

In the United States today, everyone depends on electricity. Tiny calculators and pocket radios run on batteries. Batteries produce small amounts of electricity. Your television set uses a large amount of electricity. Air conditioners and refrigerators use even more.

Large quantities of electricity travel long distances to get to your home. They travel over transmission (trăns **mĭsh′**ən) lines. In some places, transmission lines are above the ground, like these. In most cities, the lines are hidden beneath the ground.

Did you know that electricity is a form of energy and that electricity can cause magnetism? Do you think magnetism can cause electricity?

Energy and work

"Energy" and "work" are ordinary words. You use them every day. In science, these words have very special meanings. Work occurs when something is made to move. If you lift your book off the floor, you have done work. 1 That is because you made your book move. You may push all day against a brick wall. When you are finished, you will be exhausted. Still, you will have done no work. 2

Energy is the ability to do work. It may take many forms. Pushing or pulling is an example of mechanical energy. The person pushing against the wall is using mechanical energy. Notice that no work is being done. But she still is using energy.

1

2

4

3

5

A sailboat is moved by wind. It uses mechanical energy. 3 A hot-air balloon rises. It uses heat energy. 4 Chemical energy drives a rocket. 5

Conservation of energy

Energy is neither created nor destroyed. It changes from one form to another. Green plants get energy from the sun. You eat the plants. Then you have the sun's energy. The plant got the sun's energy as light and heat. The plant stored it as chemical energy. Your body will change some of it to mechanical energy. You may crank a small **generator (jĕn′ə rā′tər)**. That will change mechanical energy to electrical energy. The electricity may light a lamp. That is light energy. 1

generator:
A machine that changes mechanical energy into electrical energy.

1

Check yourself

1. Define "energy."
2. What kind of energy does a generator produce?
3. List five types of energy.

Electricity in circuits

For electricity to do work, it must be in a circuit. A circuit is a closed loop. Every electrical circuit must contain three parts. First, there must be a source of electricity. Second, there must be a resistance (rĭ **zĭs′**təns). A resistance is something that opposes the flow of electricity. The resistance is where the work is done. Finally, there must be wires, or some way to complete the circuit.

Here is a simple closed circuit. **2** The source of electricity is a battery. The resistance is a light bulb. Wires connect the resistance to the source. An electric current flows through the wires. The diagram shows the direction of the **current**. **3** The charges flowing through the wire are negative. They begin at the negative side of the battery. They flow around the circuit. Finally, they return to the positive side of the battery.

current:
Flow of electric
charges.

2

3

Key:

— dry cell

— light bulb

— wire

⊖ negative charge

→ direction of current

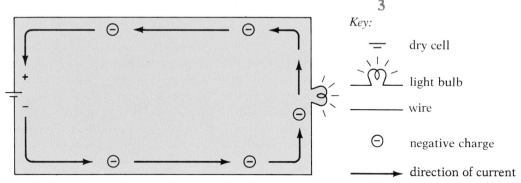

Conductors and insulators

All materials resist the flow of electricity. Some materials resist more than others. Materials with little resistance are called conductors (kən dŭk′tərz). Materials with great resistance are called insulators (ĭn′sə lā′tərz).

Activity

Working with circuits

1. You need a D-cell battery, wire, a bulb holder, and a bulb. You also need some cellophane tape.

2. Set up the equipment as shown. 1

3. Touch the two wires to different objects, such as a key, a wooden pencil, or a plastic cup. 2

4. Keep a record of what materials completed the circuit. (When the circuit is completed, the bulb will light.)

5. What always completed the circuit?

Touching the wires to a metal object will always make the bulb light. Most metals are conductors. Wood, plastic, cloth, and glass are very poor conductors. That means they are insulators.

1

2

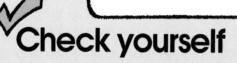

Check yourself

1. List the three parts of an electrical circuit.
2. What is the flow of electric charges called?
3. What kind of charges flow in a circuit? Where do they begin?
4. Explain the difference between a conductor and an insulator.

Series circuits

Often, there is more than one resistance in a circuit. Then there are two different ways to connect those resistances. One of the ways is in series. Here is a series circuit. 3 Notice that the electric current must pass through one resistance to get to the next. What do you think would happen if one bulb were to burn out?

3

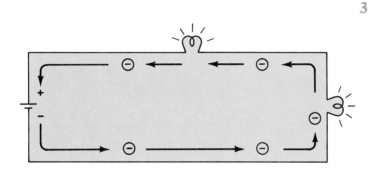

Activity

Working with a series circuit

1. You need a battery, wire, and cellophane tape. Two bulbs and bulb holders are also needed.

2. Connect the bulbs and battery in series.

3. Unscrew one bulb. What happens to the other?

4. Repeat step 3 using the other bulb.

5. What do you think the reason is for what happens?

In a series circuit, current must pass through both bulbs. If it does not, there is not a complete circuit. Either both bulbs light, or neither lights.

Batteries may also be connected in series. Then the batteries add their energies together. The bulb will be brighter in the circuit. 1

1

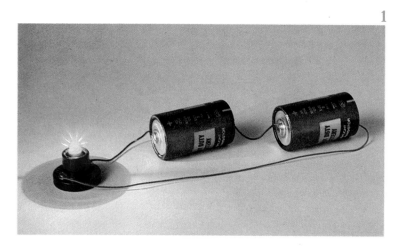

Parallel circuits

Another way to connect resistances is in parallel (**pǎr′ə lĕl′**). In the diagram, the two bulbs are connected in parallel. 2 Follow the flow of the current through each of these bulbs. What would happen if one bulb burned out?

When bulbs are in parallel, there are actually two circuits. Current can flow through either circuit. Or it can flow through both. Breaking one circuit does not affect the other. When one bulb is unscrewed, the other stays lit.

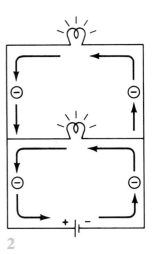

2

Check yourself

1. Name two kinds of circuits.
2. What happens when one bulb is unscrewed in a series circuit?
3. What happens when one bulb is unscrewed in a parallel circuit?

Two kinds of batteries

There are many different kinds of batteries. Any battery that contains liquid is a wet-cell battery. Automobile batteries contain wet cells. These cells are connected in series. That way the cells add their electricity together.

The liquid (water and battery acid) in a battery is called electrolyte (ĭ lĕk′trə līt′). The metal parts are called electrodes (ĭ lĕk′trōdz′). Together they generate electricity by chemical reaction.

CAUTION: Never open batteries. They contain liquids that can burn your skin. It is not safe to take batteries apart.

Not all batteries have liquid in them. The batteries used in flashlights and transistor radios, for example, are called **dry-cell batteries** (drī sĕl **bǎt'ə rēz**). This is how one type of dry-cell battery looks when cut in half. 1

Actually, a dry-cell battery is not fully dry. The electrolyte is moist. It will dry out if it is opened up to the air. But these batteries are called dry cells because they do not have any loose liquid in them.

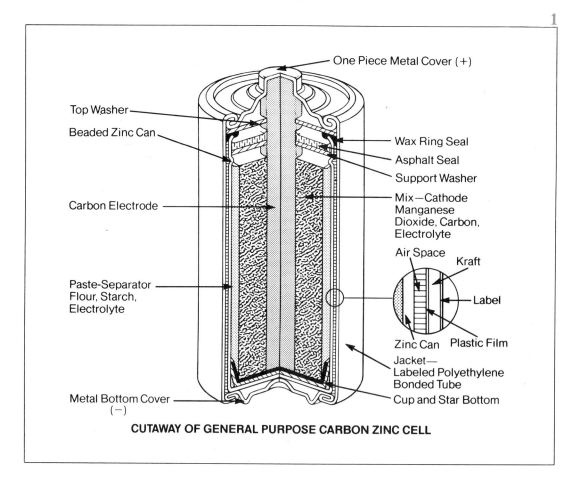

CUTAWAY OF GENERAL PURPOSE CARBON ZINC CELL

Electrical generators

Most electricity in your home does not come from batteries. It comes from a generating plant. Electrical generators change mechanical energy into electrical energy. This generator is turned by steam. 2 Energy stored in fuel is released when the fuel is burned. That energy boils water. The water changes to steam. Steam pressure then turns the generator. The end product is electrical energy for all of us.

Electricity is generated when a coil of wire moves through a magnetic field. A **magnetic field** surrounds every magnet. It is made up of lines of magnetic force. That force is able to attract certain metals. The magnetic field is strongest at the ends of the magnet.

An electrical generator contains a huge magnet. It also contains many kilometers of wire.

magnetic field:
Lines of magnetic
force that surround a
magnet.

2

Electricity and magnetism

Magnetic fields help produce electricity. Electricity flowing through wires also produces magnetism.

Dylan had a battery he wanted to test. He was not sure if it was any good. An instrument that is used to test batteries is called a **galvanometer** (găl′və **nŏm′**ĭ tər). Dylan did not have one, but he knew how to make one.

Dylan used a compass, a piece of cardboard, and some wire. He also used a pair of scissors. He folded the cardboard as shown in the drawing. That was so it would stand up. Then he cut the two notches. 1 Next, Dylan placed the compass on top of the cardboard. 2 Then he wrapped several turns of wire around the compass. 3 That completed his simple galvanometer.

galvanometer:
An instrument that detects small electric currents.

Finally, Dylan connected the galvanometer to the battery. As he did that, he watched the compass. The needle moved. What did that tell him?

Dylan decided that current was flowing in the wire. That was why the needle moved. The current in the wire caused a magnetic field. That field made the compass needle move. 4 He decided that the battery was good.

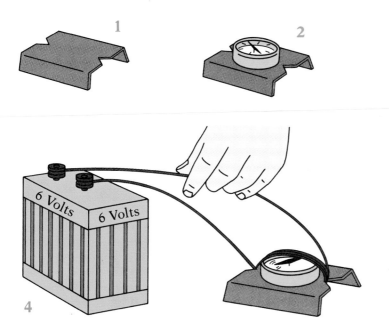

Magnets with electricity

Lee thought she would like to make a strong magnetic field with wire. She wrapped 100 turns of insulated thin wire around an iron bolt. Then she connected the ends of the wire to a battery. Look what happened. 5 Lee made a magnet that picked up paper clips and staples. It also picked up nails. When Lee disconnected one wire, all the clips and nails fell from the magnet. What did that prove?

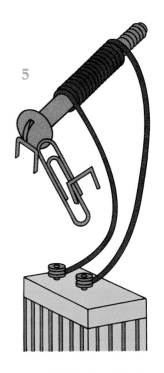

What Lee made is called an **electromagnet** (ĭ lĕk′trō **măg**′nĭt). An electromagnet is a soft iron core. It becomes a magnet while electric current flows through a coil of wire surrounding it. Electromagnets have many uses. 1 Motors contain electromagnets. Bells and buzzers use electromagnets. They change electrical energy into motion and sound.

Look at the bell. 2 Can you find the electromagnets? There are two. When current flows through the electromagnets, the striker hits the bell, and the bell rings. This kind of bell rings over and over because when the striker moves, something else happens: The striker breaks the circuit. In other words, it interrupts the flow of current. A spring pulls the striker away from the bell. That closes the circuit. Electricity flows through the magnet again and the bell rings again. This kind of bell rings hundreds of times per minute.

A buzzer works the same way as a bell. A buzzer uses a metal strip, or card, to make noise. That is the only difference between a buzzer and a bell.

Very large electromagnets are useful in industry. In automobile junkyards, they pick up great amounts of heavy scrap metal.

electromagnet: A magnet that works only when there is an electric current flowing.

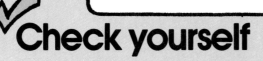

Science in our world

Bells and buzzers are common uses for electromagnets. Electromagnets are used in vibrating circuits. Electromagnets were first used for communication in 1844. That was when Samuel Morse invented the telegraph. By making clicks, messages could be sent through telegraph wires. Morse code is still in use today.

In 1875, Alexander Graham Bell invented the

telephone. Both the telephone and the telegraph use electromagnets to make sounds. Messages once took weeks to cross the country. Now they are received and answered in minutes. In some cases, they are answered in seconds.

Careers

Michael Faraday was an English scientist. 1 He is known for his discoveries in electricity. When Faraday was young he worked for a book binder. He became interested in science after reading some books at the binder's shop.

1

Faraday discovered that electricity is produced when a coil of wire moves through a magnetic field. He is given credit for inventing the first electric generator.

Faraday made many other discoveres. He showed that a current in one circuit could cause currents in another circuit to flow. Many electrical devices depend on that discovery. He also showed how a magnetic field effects light rays.

Review the main ideas

Energy cannot be created or destroyed. It changes from one form to another. Energy is the ability to do work. Work is done when something is moved.

Electricity is a form of energy. It does work when it is in electric circuits. Any electric circuit must have a source of electricity, a resistance, and a path connecting them. Electricity flows in a closed circuit from negative to positive. Batteries and generators are sources of electricity. Batteries change chemical energy into electrical energy. Generators change mechanical energy into electrical energy.

Electric current is resisted by all materials. Those that resist most are called insulators. Those that resist least are called conductors. Most metals are good conductors. Plastic, glass, wood, and cloth are good insulators.

Electrical resistances can be combined in series circuits. In a series circuit, current must flow through all the resistances. If one resistance burns out, the circuit is broken. Electrical resistances can also be connected in parallel. In a parallel circuit, there is more than one path for the current to follow. When one resistance burns out, a complete circuit may still exist for each other resistance.

Magnetism and electricity are related. In a generator, coils of wire break lines of magnetic force. That generates an electric current in the wire. When there is a current in a wire, a magnetic field forms around the wire. That field can be concentrated to form an electromagnet.

Small amounts of electric current can be detected by a galvanometer. A galvanometer can be made from a compass and wire.

Chapter review

Science vocabulary

Choose the items in column B that best match the terms in column A.

A	B
1. generator	a. a magnet's force
2. current	b. detects small currents
3. magnetic field	c. magnets and wires
4. galvanometer	d. works only while current flows through it
5. electromagnet	e. a flow of electric charge

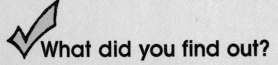 What did you find out?

Answer the following.

1. Three forms of energy are _____ , _____ , and _____ .
2. In a generator, wire passes through a _____ .
3. Unscrewing one bulb turns off all bulbs in a _____ circuit.
4. Electric current is resisted by _____ .
5. Plastic is a good electrical _____ .
6. Street lamps are usually connected in series. (True/False)

Think and write

1. Find out what a short circuit is. Tell why it is dangerous.
2. Find out how nuclear energy is used to generate electricity. What are its advantages? What are its disadvantages?

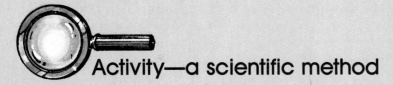

Activity—a scientific method

Purpose: to make an electrical generator.

Materials needed

compass

1 meter (about 1 yard) of insulated hook-up wire

10 meters (11 yards) of #24 enamelled wire

piece of cardboard, 8 centimeters × 8 centimeters

paper or plastic tube   scissors   bar magnet   cellophane tape

Procedure

1. First build a galvanometer. (See page 272.) Be sure to strip about 2 centimeters of insulation from the wire's ends.

2. Wind the enamelled wire around the tube. Start about 10 centimeters from the end. This is your lead. Wind until 10 centimeters of wire is left. This is your other lead.

3. Use a piece of tape to keep the wire from unwrapping.

4. Scrape the insulation from both ends of the wire.

5. Connect the ends of the coil to the galvanometer.

6. Move the magnet in and out of the tube. Record what happens.

7. Move the magnet slowly, then rapidly. Record what happens.

Observations

1. What happened when the magnet was moved through the coil?

2. How did changing the speed affect things?

3. What conclusions can you draw?

Sound

In this chapter, you will find out that

■ Sound is made when something vibrates.
■ Sound travels through liquids, solids, and gases.
■ Sound reaching our ears is heard.
■ Sound communicates many messages to us.

Before you begin

Have you ever heard silence? Silence is a complete absence of sound. Even when you are in the library, there is never complete silence. There is always some kind of background sound.

How do you use sound? Have you ever considered how important it is to you? Could you order lunch without making a sound? Would it be easy? Could you make a telephone call without a sound? Why do you use a telephone? Why do you listen to a radio or to a record player? Do you like the sounds they make?

Have you ever gone to see an orchestra? Or perhaps you went to hear one. Orchestras make sounds. They make sounds that are pleasant to our ears. What is sound? How does it travel from its source to your ear? How are sounds made? Did you know that sound is made when something vibrates? Did you know that sound travels through liquids, solids, and gases? Sound cannot travel through a vacuum. Can light?

Stop and listen

Close your eyes. Listen. 1 How many different sounds can you hear? Do you hear the sound of people's voices? Perhaps you hear traffic noises. Maybe you hear birds singing or music playing. When you start listening, you might realize that it is not very often that your surroundings are silent.

We depend on our ears to tell us about our surroundings. Can you tell where all the sounds you are hearing come from? Are they pleasant or unpleasant to hear?

How are sounds made?

Luis is in the school play. He is excited about his role as announcer. He is also responsible for letting the audience know when it is time to return to their seats after intermission. Luis does this by clashing cymbals (**sĭm′**bəlz) together at the beginning of each act. 2

1

2

Even hitting the cymbals once makes a loud sound. And it lasts a long time. When Luis looked at the cymbals closely, he saw that the metal was constantly shaking back and forth.

When an object vibrates, it moves back and forth repeatedly. Its back-and-forth movement moves the air around it. As the object moves outward, it pushes air away. As the object moves back in, air rushes in to fill the empty space. 3 The air next to

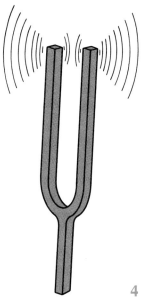

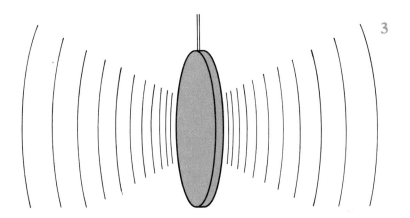

the object passes the vibration to the air next to it. After a while, the vibrations are passed along farther and farther from the object. When vibrations strike our ear, we hear them. We call the vibrations sound waves (sound wāvz).

Sometimes, you can see the vibrations. Have you ever used a tuning fork? 4 It is a small, steel tool with two prongs. A tuning fork is used by musicians to tune instruments. Striking it against a hard rubber object causes it to vibrate. Its vibrations produce sound. If you hold it near your ear, you can hear a musical tone. You can tell the tuning fork is vibrating by placing a vibrating one in a pan of water. The vibrations from the tuning fork cause ripples, or waves, in the water. 5

Perhaps you have thrown a pebble into a pond. Do you remember the rings of water that moved outward? 1 They spread slowly from where the pebble hit the water. The rings fade as they move farther away from the pebble. In the same way, sound waves fade as they move farther away from the source of the vibrations.

1

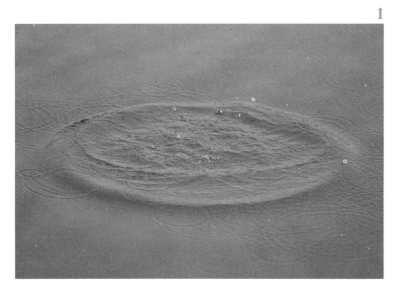

✓ Check yourself

1. What is the back-and-forth movement of an object called?
2. What does this movement do to air?
3. How does sound travel in air?
4. Define "sound waves."

How do we hear?

Imagine that you were nearby when a tree crashed onto the forest floor. How would you hear it? You would use your ears of course.

The outer part of your ear acts like a funnel that catches sound. Once the sound waves are caught, they go through a little tube, or **canal**. Then the sound waves hit a very delicate part of your body, your eardrum. 2 The eardrum is a sensitive membrane. It picks up the vibrations of sound waves. But that is not the whole story of hearing.

The eardrum then passes the vibrations along to three tiny bones that are connected to it. Then the vibrations hit a part called the **cochlea (kŏk′lē ə)**. The cochlea looks like a snail, and it is filled with liquid. 3 Vibrations cause the liquid to move around. This movement triggers a nerve connected to the cochlea. The nerve sends a message to your brain. The result is that you hear.

canal:
A passageway.

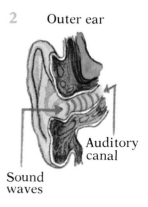

2   Outer ear

Sound waves

Auditory canal

cochlea:
The inner ear,
containing a liquid.

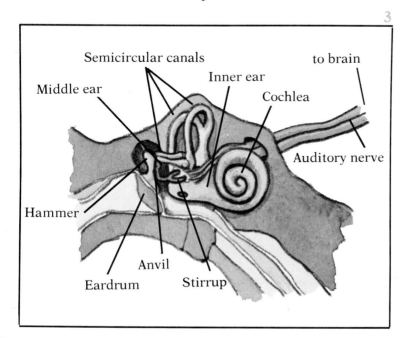

3

Semicircular canals   to brain

Inner ear

Middle ear   Cochlea

Hammer

Auditory nerve

Anvil

Eardrum   Stirrup

285

Check yourself

1. What do sound waves hit after passing through the ear canal?
2. List the three tiny bones connected to the eardrum.
3. What do we call the part of the inner ear that contains liquid?

The speed of sound

Which travels faster, sound or light? Have you ever been to a fireworks display? 1 Do you hear the sound of the explosion at the same time you see the burst of light? The noise is produced at the same time as the flash of light. But because light travels much faster than sound, we see the flash before we hear the boom.

1

The speed of sound in air is about 330 meters (about 1100 feet) per second. That means that sound travels 330 meters in 1 second. If you count the seconds between the flash and the sound, you can find how far away the explosion was. Suppose you see an explosion and count 4 seconds until you hear the sound. Then the explosion was 4 × 330 meters, or 1320 meters, away from you.

Sound travels through many solids and liquids. The speed at which sound travels changes. Its speed depends on what it is traveling through.

Have you ever used a tin-can telephone? 2 Your voice travels through the wire to your friend's ear. The sound travels much more quickly through the wire than it would through the air. Sound travels about 330 meters per second in air. It travels about 5000 meters (16,500 feet) per second in the steel wire! Therefore, sound waves travel more quickly through solids than through gases and liquids.

✓ Check yourself

1. Which travels faster, sound or light?
2. How fast does sound travel through air?
3. Does sound travel faster through solids or gases?
4. If you count the seconds between the flash and the sound of fireworks, what will you find out?

Different qualities of sound

How would you describe the sound of a siren? Maybe the first word that comes to mind is "loud." A siren is meant to be heard. Some sounds, like a whisper, are soft. They are not meant to be heard over a great distance. The quality of loudness or softness is called **volume**. You are probably familiar with the volume control knob on a TV or stereo. 1 Volume is just one quality of sound.

volume:
The loudness or softness of a sound.

The two instruments in the photograph make very different sounds. 2 The highness or lowness of a sound is its **pitch**. A flute can produce high-pitched sounds. A tuba can produce low-pitched sounds. Pitch is another quality of sound. The sounds around you differ in their pitch and volume.

pitch:
The highness or lowness of a sound.

2

A tape recorder captures and records sound waves. Using one can help you understand how much sound varies.

Activity

Recording sounds on tape

1. You need a tape recorder and a blank cassette.
2. Learn all the controls on the tape recorder. Learn how to record, stop, play, and rewind. Find the volume control. You should always record with volume at a low setting.
3. Make a 5-minute tape. Record some interesting sounds that come from different sources. Remember, too many sounds at once can be confusing.
4. Play back your tape for your class. Can they identify the sounds? Were there enough different sounds to keep their interest?

Lynn recorded several sounds on her tape. It started off loud and noisy. She recorded in the cafeteria at lunchtime! The sound of students talking, chairs scraping, and trays clanging made her glad she recorded for only 30 seconds.

Next, she recorded her brother playing the piano. What a difference! This was not noise. It was music—a very pleasant sound. Then Lynn recorded her 5-month old brother. Perhaps she will keep this tape and play it for her brother when he is older.

✓ Check yourself

1. Name two qualities of sound.
2. What is the difference between volume and pitch?
3. Does a tuba or a flute have a higher pitch?
4. Name a device that records sound waves.

Sound as communication

You probably never stop to think about it, but almost every sound you hear communicates (kə **myoo'**nĭ kāts') something to you. "Communicates" means "conveys or sends meaning." When you hear a sound, your brain interprets it, or "figures it out." That sound has communicated something to you.

The most important means of communication is **verbal communication** (**vûr'**bəl kə myoo'nĭ-kā'shən). 1 Verbal communication is communi-

verbal communication: Communication using words.

1

291

cation that depends on the use of words. Not all communication in words produces sounds. If you are reading this paragraph silently to yourself, the words on the page are communicating to you although no sound is produced. People who use sign language communicate without sound. **1**

1

We can receive sounds through nonverbal communication (nŏn **vûr′**bəl kə myo͞o′nĭ **kā′**shən) as well. Think of these sounds: a fire-alarm bell, a train whistle, an ambulance siren. Each of these sounds gives you a message even though the message is not in words.

Another way sounds communicate is by calling up certain emotions in us. Little puppies are restless unless they can hear the sound of their mother's heartbeat. If you place a ticking clock under a blanket in the puppy's box, however, the puppy will think the clock is its mother's heartbeat. The clock will make the puppy calm down. **2**

Obviously, the sound is communicating something to the puppy. Can you think of any sounds that might communicate to a person's emotions? What about laughter? What about a baby's cry?

2

In nature, the song of a bird, the chirp of a cricket, the patter of raindrops are pleasing to hear. 3   Of course, sound can communicate with us in unpleasing ways, too. We call unpleasant sounds **noise** (noiz). Noise can be a very loud or sudden sound. Noise is sound without a pattern, which is confusing to our ears.

noise:
Unpleasant sounds.

3

 Activity

Sounds and noise

What sounds are pleasant to you? Which are just noise?

1. See if you can describe several sounds.

2. Use a chart like the one you see on the page. It shows an example of what Ed thought was an unpleasant sound. Do you agree with him?

3. Think of five more sounds. Look around your classroom or at home for sources of familiar sounds. Thumb through the pages of your book for ideas. You may want to cut out magazine pictures to show your sources of sound.

4. Copy this chart and fill it in.

Source of Sound	Pleasant	Unpleasant
Squeaky chalk		

✔ Check yourself

1. How is a sound communicated to you?

2. What is verbal communication?

3. How can words be communicated without sound?

4. Name a language that does not use sound.

5. Give examples of nonverbal sounds.

6. Give an example of how sound can communicate feelings.

Science in our world

The world around you is full of sounds that help people communicate with each other. Music is one way of communicating. Music is an important part of all cultures. Sometimes, music is loud. By listening to very loud music over a long time you can damage your ears. Your hearing is an important way for you to know what is going on around you. Protect your ears by avoiding loud sounds.

Some people have ear injuries or other conditions that cause poor hearing. Many people wear hearing aids to help them hear, just as many people wear eyeglasses to help them see. Not all hearing problems can be helped with hearing aids, however. Those who cannot hear at all live in a soundless world. But these people learn to communicate in other meaningful ways.

Careers

Otologists (ō tăl′ə jĭsts) are medical doctors who specialize in the care of the ear. Serious damage to the ear can make a person deaf (dĕf). Otologists help prevent loss of hearing. They also are able to cure or help some deaf people. They perform ear operations. They also prescribe hearing aids.

Otologists examine ears with an otoscope, a device with a light that helps the doctor see inside ears. 1 Sometimes, otologists use an audiometer.

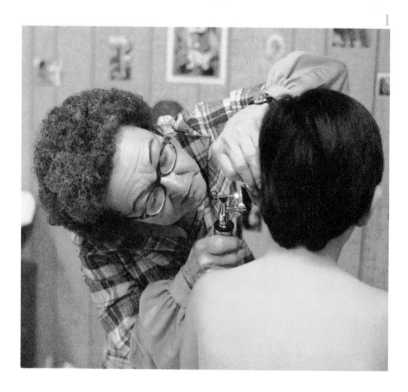

This is a machine that makes musical tones. It helps the doctor know how well you hear.

Otologists go to college for 4 years and then to medical school. It takes a long time, and a lot of work, to become an otologist.

Review the main ideas

Our lives are filled with different sounds. Sounds are caused by vibrations that travel through the air as sound waves. We catch sound waves with our outer ears. The vibrations strike our eardrums, and eventually a message of sound is sent to our brains. Sound travels more slowly than light, but it still travels at 330 meters per second in air. In water or steel, sound travels even faster.

Sounds differ from one another a great deal. They have different pitches. Some are high and some are low. They also have different volumes or loudness. Some sounds are pleasing and others are just noise. Different sounds can even communicate emotions to us.

Chapter review

Science vocabulary

Choose the items in column B that best match the terms in column A.

A   B

1. vibrate   a. snail-shaped part of ear

2. noise   b. unpleasant, confusing sound

3. eardrum   c. move back and forth rapidly

4. volume   d. sensitive membrane

5. cochlea   e. loudness

6. pitch   f. the highness or lowness of sound

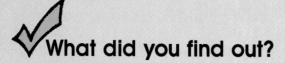 What did you find out?

Answer the following.

1. Vibrations travel in the form of sound _____ .

2. After sounds enter the ear canal, they strike the _____ .

3. The cochlea contains a _____ that carries vibrations.

4. _____ travels faster than sound.

5. Sound travels _____ through solids than through gases.

6. Music and words are ways in which people _____ .

Think and write

1. Describe, in order, the parts of your ear that help you to hear.

2. Is there a difference between noise and music? Explain.

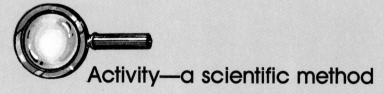 Activity—a scientific method

Purpose: to find out if hearing is better with two ears or one.

Materials needed

2 coins

Procedure

1. Work with a partner.

2. Close both your eyes.

3. Your partner will click the two coins together. The coins will click in front of you and behind you. They will click above your head and beside you.

4. Each time you hear a click, point to where it comes from. Your partner will record whether you are right or wrong.

5. Your partner should click from at least eight different places.

6. Cover one ear. Keep your eyes closed. Have your partner repeat steps 3 through 5.

7. Record what you have discovered.

8. Change places with your partner. Repeat the activity.

Observations

1. How many times were you right about where the click was?

2. Were you right more often with one ear or with two?

3. Why do you think you got the result you got in step 3?

4. Which clicks were hardest to locate?

19

Light and lenses

In this chapter, you will find out that

- Light travels at different speeds in different materials.
- When light changes speed, its path bends.
- Different shaped lenses bend light differently.
- Telescopes and microscopes use combinations of lenses.

Before you begin

If scientists studied only things that could be seen with the naked eye, we would know much less about the universe than we do. Some things are too small or too far away to be seen with the naked eye. Scientists invented different sorts of lenses to see those things.

If everybody had to live with the eyesight with which they were born, this country would be a very dangerous place. More than half the drivers in America wear lenses to correct their eyesight. How do you think traffic safety would be affected if there were no lenses for eyeglasses?

Do you know how lenses work? Did you know that they bend light? Think of how many uses there are for lenses. How are lenses used at a ballpark? How are they used by a doctor? Why does a photographer need lenses?

Look around your classroom. Do you see any lenses?

301

Light in motion

Have you ever heard the expression "with the speed of light?" Light moving through air always travels at one rate of speed. It travels about 299,700 kilometers (about 186,000 miles) per second. Light always travels in straight lines through air.

When light travels through materials other than air, its rate of speed changes. It speeds up, or it slows down. When the speed of light changes, the path the light travels changes, too. Instead of traveling in straight lines, light bends from its original path. The bending of light is called **refraction** (rĭ **frăk′**shən). 1

refraction:
The bending of light.

1

Have you ever noticed your feet as you stood in shallow water? They looked as though they were not connected properly to your legs. Your feet looked strange because the light above the water was moving at a different rate of speed than the light below the water. Water refracts light.

Activity

Studying refraction

1. You need a clear cup, a ruler, and water.
2. Fill the cup about three-fourths full of water.
3. Stand the ruler in the cup.
4. Look at the ruler in the cup from the side.
5. Watch the part of the ruler below the water. While you watch, back away from the cup. What do you notice?

The ruler in the cup looks broken. That is because the speed of light changes from air to water. Below the water, the ruler seems to bend. 2 It looks wider at the wide part of the cup and narrower at the narrow part. The cup of water acts like a lens.

2

Lenses

Scientists make use of refraction to help people. **Lenses** (lĕn′zəz) are transparent, or see-through, objects designed by scientists. Lenses bend light in different ways. The more curved the lens is, the more it will bend light. The less curved the lens is, the less it will bend light. The most common use of lenses is in eyeglasses. When the eye does not see properly, a lens can be designed to bend the light to improve vision.

Some lenses are flat on one side and curved on the other. Some are thick in the middle and thin at the ends. These are called **convex** (kŏn′vĕks′) lenses 1 Other lenses are thin in the middle and thick at the ends. They are called **concave** (kŏn′kāv′) lenses. 2

lenses:
Transparent objects that bend light.

convex lens:
A lens that is thick in the middle and thin at the ends.

concave lens:
A lens that is thin in the middle and thick at the ends.

1
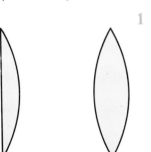

2

Refracting telescopes

Convex lenses **invert** (ĭn **vûrt′**) images. That means they turn them upside down. If you have ever put slides into a slide projector, you know you have to turn them upside down. Then the lens in the projector will invert the images on the slides and show them right side up. By putting a second lens in front of the first, scientists learned that they could invert the image twice. The first lens turned the image upside down. The second lens inverted the upside-down image, and turned it right side up.

The two-lens system in telescopes makes large images by refracting, bending, light. Telescopes with convex objective lenses are called **refracting** (rĭ **frăkt′**ing) **telescopes.** 2   The lens near the eye is the **eyepiece** (ī′pēs′) **lens**. The lens that is farthest from the eye is the **objective** (əb **jĕk′**tĭv) **lens**. The objective lens collects light. The more light it collects, the brighter and clearer the image becomes.

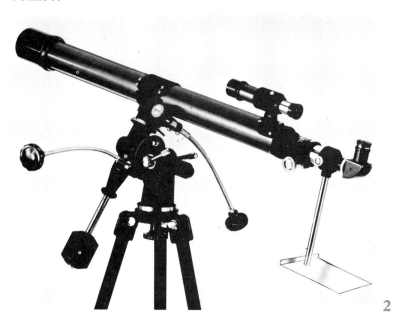

2

invert:
To turn upside down, or reverse in position.

refracting telescope:
A telescope that uses lenses to collect light.

eyepiece lens:
The lens of a telescope that is near the eye.

objective lens:
The lens of a telescope that is farthest from the eye.

✓ Check yourself

1. Define "lens."
2. Name the type of lens that is thick in the middle and thin at the ends.
3. Identify the lens in a telescope that is nearest the eye.
4. Name the lens in a telescope that is farthest from the eye. What is the purpose of this lens?
5. How does the amount of light the objective lens collects change the image?

Reflecting telescopes

Instead of using lenses to collect light, some telescopes use mirrors. 1 Such a telescope is a **reflecting** (rĭ flĕkt'ing) **telescope.**

Mirrors do not reverse images from right side up to upside down. They reverse images from left to right and from right to left. Have you ever noticed that when you stand before a mirror and raise your right hand, the image in the mirror seems to show the left hand raised?

Just as it takes two lenses to create a correct image, it takes two mirrors to create a correct image. Reflecting telescopes use this principle. The first mirror in the telescope reverses the image from left to right. The second mirror in the telescope reverses that image and presents it correctly to the eyepiece.

reflecting telescope: A telescope that uses a mirror to collect light.

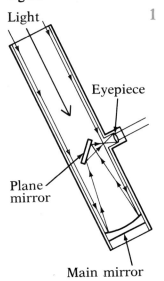

306

Activity

Finding out about telescopes

1. You need reference materials about telescopes.

2. Find out about the huge telescopes that astronomers use.

3. Why must special buildings be built just to hold these telescopes?

4. Find out how these telescopes work. Where are they?

5. Why do scientists expect better results from smaller telescopes in space?

6. Collect pictures and draw a diagram of one of these telescopes. Make a collage, a poster, or an illustrated report of your findings.

These giant telescopes are built on mountain tops. The top half of the building turns as the Earth turns. That way the telescope can stay aimed at one star or planet. 2

Even on a high mountain, the Earth's atmosphere blurs the telescope's view. Small telescopes in space do not have to look through air. That makes their pictures clearer than those of telescopes on Earth.

Film may be placed in telescopes so that they can be used as giant cameras. While astronomers view certain areas of the sky, the telescope is photographing those areas.

2

✓ Check yourself

1. What type of telescope uses mirrors to collect light?
2. Why does a reflecting telescope use two mirrors?
3. Where are giant telescopes built?

Binoculars

Have you ever seen **binoculars** (bə nŏk′yə lərz)? **1** Binoculars are hand-held objects. They allow us to see things at a greater distance than usual. Binoculars use prisms (**prĭz′əmz**) in much the same way reflecting telescopes use mirrors. Each side of a pair of binoculars contains four

binoculars:
A hand-held optical device that can be looked through with both eyes.

1

reflecting faces. The diagram shows how the reflections work in binoculars. 2

The prisms in binoculars are there for a good reason. They allow the objective lenses to gather light from a wide area. Then they bring the light to the narrow distance between your eyes. Notice how the objective lenses are farther apart than the eyepieces. 1, 2

Binoculars, like telescopes, need to collect light. Small objective lenses do not collect as much light as large objective lenses do. Small objective lenses, however, may magnify as much as big ones. For this reason, each pair of binoculars is marked with two numerals. One numeral tells how many times the pair of binoculars magnifies the object. The other numeral tells the diameter of the objective lens.

The binoculars in the photograph are marked 6 × 30. The first numeral means they make things look six times closer than they are. The second marking means that the diameter of the objective lens is 30 millimeters. The larger the diameter of the objective lens, the more light the lens collects and the brighter the image.

2

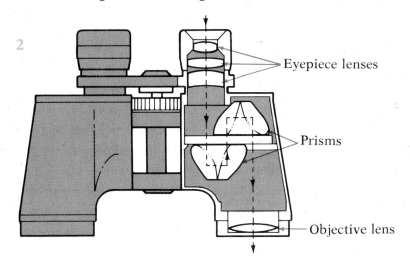

Eyepiece lenses

Prisms

Objective lens

✓ Check yourself

1. What do binoculars use to reflect light?
2. What do binoculars use to gather light?
3. Do small objective lenses collect more or less light than large lenses?
4. Which numeral on a pair of binoculars tells how many times the binoculars magnify an object?

Hand lenses and compound microscopes

Telescopes and binoculars are useful for looking at faraway things. There are also instruments used for looking at things that are near but very, very small.

Microscopes are instruments that are used to view things too small to be seen with the naked eye. One kind of microscope is called a **simple microscope** (sĭm′pəl mī′krə skōp′). A simple microscope is a single magnifying glass, or hand lens. 1 Hand lenses are most useful for magnifying things. Hand lenses also gather light. They can focus light rays to produce a burning heat.

Scientists in a laboratory use a microscope that is called a **compound microscope** (kŏm′pound′ mī′krə skōp′). 2 Compound microscopes contain two or more magnifying lenses. They can magnify things as much as 1000 times or more.

1

**simple microscope:
A hand magnifier
using one lens.**

**compound
microscope:
An instrument that
uses 2 or more lenses.**

310

An object to be viewed through a compound microscope is placed on a slide. The slide is placed on a platform below the objective lens. A small, movable mirror below the platform is used to collect light and to direct it onto the slide. Sometimes, a compound microscope may have two or

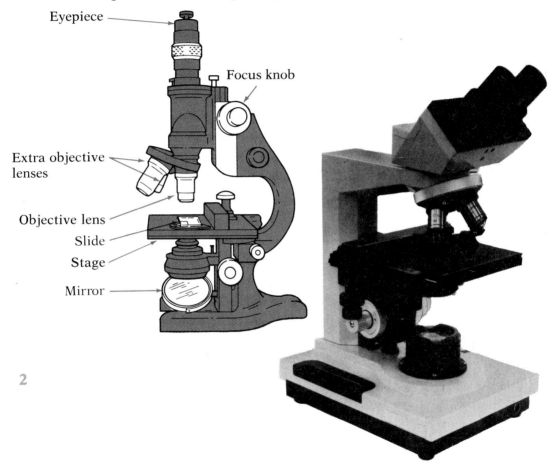

Eyepiece

Focus knob

Extra objective lenses

Objective lens

Slide

Stage

Mirror

2

three objectives. These objectives have different magnifying powers. They can be swung into position above the slide when needed. Most compound microscopes have single eyepieces for one eye viewing. Some kinds have two eyepieces so both eyes can be used at the same time.

✓ Check yourself

1. Name an instrument that uses lenses to make small things appear larger.
2. What is a microscope that uses two or more lenses called?
3. Where is an object placed when viewed through a compound microscope?

Science in our world

With the invention of the compound microscope, scientists found many tiny living things that had never been known to exist. These things are so small that they can be seen only with a microscope. They are called **microorganisms** (mī′krō ôr′gə nĭz′əmz). Some microorganisms cause illness in people.

Thanks to the invention of the compound microscope, these microorganisms can be seen and steps can be taken to control their growth. Many medicines have been developed to control and destroy certain disease-causing microorganisms. Wonder drugs such as **penicillin** (pĕn′ĭ sĭl′ən) and other **antibiotics** (ăn′tē bī ŏt′ĭks) have been discovered only in the last 50 years. These discoveries would not have been possible without the aid of the compound microscope.

microorganism: Tiny living things that can be seen only with a microscope.

penicillin: An antibiotic.

antibiotics: A group of wonder drugs that fight certain microorganisms that cause disease in people.

312

Careers

Ophthalmic laboratory technicians (ŏf thăl′mĭk lăb′rə tor′ē tĕk **nĭsh′**ənz) grind and finish lenses. **1** They are also known as optical mechanics. Most of their work is done in preparing lenses to correct vision problems. They start with mass-produced lens blanks. Then they grind and polish the lenses. They measure carefully to make sure the lenses fit the prescriptions.

Optical mechanics make lenses for other purposes, too. Telescopes, microscopes, and binoculars all require lenses that are made with great care.

1

Review the main ideas

Light traveling through air travels in a straight line. But as light moves from one material to another, the line along which it travels bends. Scientists have learned how to use what they know about bending light to make lenses that will help them investigate the universe.

Tools useful for seeing objects that are far away include telescopes and binoculars. Telescopes are of two kinds. Refracting telescopes use convex objective lenses to gather light. Reflecting telescopes use mirrors to gather light. Binoculars use prisms and lenses. Convex objective lenses are always used in pairs to correct the way such lenses invert images. Reflecting faces in telescopes and binoculars are always used in pairs also.

Microscopes are tools used for seeing things too small to be seen by the naked eye. Simple microscopes are single magnifying glasses, or hand lenses. Compound microscopes are instruments that make use of two or more magnifying lenses.

Chapter review

Science vocabulary

Choose the items in column B that best match the terms in column A.

A

1. refraction
2. invert
3. refracting telescope
4. reflecting telescope
5. simple microscope
6. compound microscope

B

a. turn upside down
b. a hand-held magnifier
c. uses a mirror to collect light
d. the bending of light
e. uses a lens to collect light
f. uses 2 or more lenses

✓ What did you find out?

Write **T** for each sentence that is true.
Write **NT** for each sentence that is not true.

1. Light moving through air travels in a straight line.
2. A lens is a transparent object that bends light.
3. In a telescope, the lens nearer the eye is called the objective lens.
4. Binoculars use mirrors to help gather light.
5. Convex lenses turn images right-side up.

Think and write

1. How might telescopes help us find out about other planets?
2. Do you think the telescope or the compound microscope is a more important invention? Why?

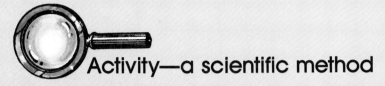

Activity—a scientific method

Purpose: to make a real image.

Materials needed

lamp with 100-watt bulb   hand lens   white onion skin paper

Procedure

1. Stand with the hand lens between the bulb and the paper. Have a partner hold the paper.

2. Hold the lens about 1 meter (3 feet) from the lamp. The paper's center should be as high as the lens.

3. Have your partner back away from the lens. The paper should stay at the same height. Your partner should keep backing up until there is an image on the paper. The image should be that of the bulb.

4. If no image forms, move the lens farther from the lamp. Then try again.

5. Exchange places with your partner and repeat steps 1–4. Record what you see.

6. Aim the lens out the window. Keep the lens between the window and the paper. Try to form an image on the paper. Record what you see.

Observations

1. What did you notice about the two images?

2. What did you notice about the image from the window?

3. What kind of lens did you use?

20

Time and distance

In this chapter, you will find out that

▪ **Average speed is the total distance traveled divided by total time.**

▪ **A distance-versus-time graph of average speed is a straight line.**

▪ **People use average speed to make travel schedules.**

Before you begin

Speed plays an important part in many people's lives. In fact, it plays an important part in everyone's life. If you are buying fresh vegetables at the market, you want to know that the vegetables got there quickly after they were picked. If the vegetables took a long time to get to the market, they would be spoiled.

People used to travel all over this country by train. Now, if they wish to go long distances, they take an airplane. Why do you think that travelers have switched from train to airplane? In Europe and Japan, new high-speed trains are in service. These trains can go at speeds of 300 kilometers (180 miles) per hour. Do you think that these trains may take business away from the airlines?

Racing cars can travel at over 320 kilometers (200 miles) per hour. This car is on one of the most famous race tracks in the world. It is the track in Indianapolis, Indiana. That is where the *Indianapolis 500* race is held each year. For race car drivers, speed is a way of life.

Measuring distance

A car enters the New Jersey Turnpike near Wilmington, Delaware. The officer on duty in the toll booth hands the driver a card. The time is stamped on the card. The driver comes to the end of the turnpike, near New York City. The officer on duty there collects the toll.

The length of the turnpike is 189 **kilometers** (about 118 miles). A kilometer is a measure of distance that is equal to 1000 meters. Suppose the first time stamped on the card was 11:00 A.M. and the second time was 3:00 P.M. How many kilometers did the driver travel in 4 hours? How many kilometers did the driver travel in 1 hour?

The speed limit is 88 kilometers (55 miles) per hour. Did the driver go faster than the limit? How can you tell?

kilometer:
1000 meters. A unit of distance.

Average speed

Look at the **speedometer** (spē dŏm'ĭ tər) of a car while it is moving. A speedometer tells the **speed**, or how fast the car is going at that moment. Speed is distance divided by time.

For example, the speedometer may read 80 kilometers (or 50 miles) per hour. 1 But drivers do not go at the same speed all the time. And they often stop along the way. Even on a turnpike, they may stop for gasoline or meals, and to look at the view.

To know the speed for a whole trip, you have to find the average. If you know how far the car has traveled and the distance, you can find its average speed. Average speed equals the total distance divided by the total time. You can find the average speed of a bouncing ball in the following activity.

speedometer:
An instrument in a car that tells how fast the car is going.
speed:
A measure of how far something travels per unit of time.

1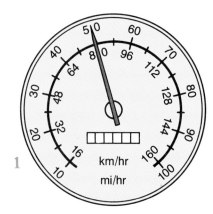

318

Activity

Finding average speed

1. You need a piece of paper, tape, a tennis ball, a clock or watch with a second hand, and a meterstick.

2. Tape the paper to the wall at a height of 1 meter (4 feet). 2   This is the height from which you will bounce the ball.

3. Have a classmate watch the clock and tell you when to start and stop bouncing the ball.

4. Bounce the ball in place the way a basketball player dribbles. Bounce it to the 1-meter mark. Bounce it for 1 minute. Count the number of bounces.

5. How many times did you bounce the ball in 1 minute? How far did it travel up and down on each bounce?

6. How far did the ball bounce in 1 minute? What was the ball's average speed?

2

When you bounce a ball from a height of 1 meter, it travels for a total distance of 2 meters on each bounce, 1 meter down and 1 meter up.  Don found that he could bounce a tennis ball 76 times in 1 minute. The ball traveled 152 meters (about 167 yards) in a minute. The average speed of the ball was 152 meters per minute.

1

Check yourself

1. What is distance divided by time called?

2. Name an instrument in a car that measures speed.

3. What is average speed?

4. If a car travels 150 kilometers in three hours, what is its average speed?

Picturing speed

Celeste bounced a ball from a height of 1 meter. She bounced it 150 times in 100 seconds. Here is her table of data.

Time in Seconds	Number of Bounces	Distance Traveled in Meters
10	15	30
20	30	60
30	45	90
40	60	120
50	75	150
60	90	180
70	105	210
80	120	240
90	135	270
100	150	300

2

Activity

Making a distance-versus-time graph

1. Use the table of the ball's speed. 2 Make a graph like the one shown. The first two points have been placed for you. 3

3

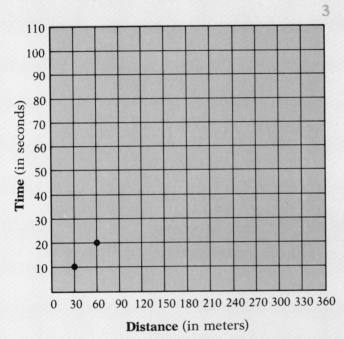

Distance (in meters)

2. After you have drawn in all the points, connect them with a line. What do you notice about the line?

You have made a distance-versus-time graph of the average speed of the ball. Ideally, this kind of graph is always a straight line.

Being on time

Often, it is important to be somewhere on time. You want to be at the theater when a movie starts. A doctor wants to be at the hospital on time. An office worker wants to be in the office on time. A business person flying to a meeting out of town wants to be there on time.

Carlos wants to know how many steps it takes for him to walk to school. It takes him 10 minutes to walk to school. He lives 500 meters (about 1650 feet) from school. By dividing, he found that he walks at an average speed of 50 meters (about 165 feet) per minute. Anna used a ruler to measure the length of Carlos's usual step. 1 It was ½ meter (1.65 feet). Carlos divided ½ meter into 500 meters. He found he had to take 1000 steps to get to school.

Think about a job in which you might be interested. Would you have to do a lot of traveling? How important are time and distance in this job?

Schedules

Bus, train, and plane schedules give the leaving time, or time of departure. They also tell the arriving time. The transportation companies know which times to print in their schedules. They know the average speeds of the buses, trains, and planes which they operate.

Look at this schedule for trains between New York City and Mount Vernon. 2   The schedule tells you when a train leaves and when it arrives. The trains run mornings, afternoons, and evenings. Note that the afternoon times are printed in a darker type. The distance between the two places is about 32 kilometers (about 20 miles).

How can you find the average speed of a train that goes from New York to Mount Vernon on Saturday? What facts would you need to know? How would you go about getting them?

NEW YORK TO MOUNT VERNON			
MONDAY TO FRIDAY, EXCEPT HOLIDAYS			
Leave	Arrive	Leave	Arrive
New York	Mount Vernon	New York	Mount Vernon
AM	AM	PM	PM
12:10	12:36	**3:40**	**4:06**
1:30	1:56	**4:10**	**4:36**
5:40	6:06	**4:30**	**4:56**
6:40	7:06	**4:54**	**5:20**
7:10	7:36	**5:18**	**5:44**
7:40	8:06	**5:38**	**6:04**
8:10	8:36	**6:02**	**6:27**
8:40	9:06	**6:19**	**6:45**
9:10	9:36	**6:40**	**7:06**
9:40	10:06	**7:10**	**7:36**
10:10	10:36	**7:40**	**8:06**
10:40	11:06	**8:10**	**8:36**
11:10	11:36	**8:40**	**9:06**
11:40	**12:06**	**9:10**	**9:36**
12:10	**12:36**	**9:40**	**10:06**
12:40	**1:06**	**10:10**	**10:36**
1:10	**1:36**	**10:40**	**11:06**
1:40	**2:06**	**11:25**	**11:51**
2:10	**2:36**	12:10	12:36
2:40	**3:06**	1:30	1:56
3:10	**3:36**	....	....
PM	**PM**	AM	AM
SATURDAY, SUNDAY & HOLIDAYS			
AM	AM	PM	PM
12:10	12:36	**4:40**	**5:06**
1:30	1:56	**5:10**	**5:36**
5:40	6:06	**5:40**	**6:06**
6:40	7:06	**6:10**	**6:36**
7:40	8:06	**6:40**	**7:06**
8:40	9:06	**7:40**	**8:06**
9:40	10:06	**8:40**	**9:06**
10:40	11:06	**9:40**	**10:06**
11:40	**12:06**	**10:40**	**11:06**
12:40	**1:06**	**11:25**	**11:51**
1:40	**2:06**	12:10	12:36
2:40	**3:06**	1:30	1:56
3:40	**4:06**	....	....
PM	**PM**	AM	AM

2

Check yourself

1. Describe the shape of a graph that shows distance–versus–time of average speed.

2. What is the importance of bus, train, and plane schedules?

3. What do you need to know to make up a schedule?

4. If a train travels at 80 kilometers an hour, how long will it take to travel 240 kilometers?

Science in our world

Speeds are measured in many athletic events. 1 Running, swimming, skiing, and ice-skating races are timed. The speeds are reported in seconds.

Almost all speeds that people use are average speeds. Racing cars go faster or slower as they round the course. Some of the fastest automobiles are run on the Bonneville Salt Flats in Utah. The cars run over a 1.6-kilometer (about 1 mile) course. First they run in one direction. Then they run the other way. The two speeds are then averaged.

Police officers sometimes use electric timers to find speeding automobiles. A timer is started when the car passes over an electric switch. The timer is stopped when the car passes over another switch at a measured distance from the first. If the time is too short, the police officer knows the car was traveling too fast. The officer can use the time to figure the average speed of the car.   1

2

Careers

To make sure that time—and money—are not wasted, businesses hire **time-study engineers**. They study working systems and people's working habits. Time-study engineers take careful notes, analyze the data they collect, and then set up new working methods. 2 The new systems suggest to employers how time can be used more wisely. Then a company may make a bigger profit or sell its products or services for less money.

Time-study engineers must know mathematics. They earn their college degrees in engineering.

Review the main ideas

Average speed is found by dividing the distance traveled by the time of travel. Distance and time for a bouncing ball can be shown on a graph. Ideally, the line on the graph will be a straight line.

Average speed is used by people who make travel schedules. Transportation companies, athletes, and police officers are some people who make use of average speed.

Chapter review

Science vocabulary

Choose the items in column B that best match the terms in column A.

A

1. speed
2. kilometer
3. speedometer
4. average speed

B

a. equal to 1000 meters
b. distance traveled divided by time
c. total distance traveled divided by total time
d. shows speed while car is moving

✓ What did you find out?

Answer the following.

1. A bouncing ball (does/does not) always move at the same speed.
2. Train schedules give _____ and _____ times.
3. If you drop a ball from a height of 1 meter, it travels a distance of _____ on each bounce.
4. If a plane travels at an average speed of 600 kilometers per hour, it (does/does not) cover 10 kilometers each minute.
5. If you travel ½ meter in one step and it takes 800 steps to get to school, you live _____ meters from school.

Think and write

1. Which travels faster, light or sound? Find out the speeds of both. Give an example of seeing something before you hear it.

2. The flying time from New York to Los Angeles is about 5 hours, 25 minutes. Flying time from Los Angeles to New York is about 5 hours. If the plane makes no stops, what makes it go faster on the return trip? Use the library to help you find out.

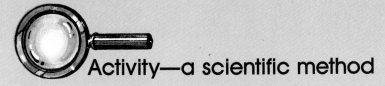

Activity—a scientific method

Purpose: to measure speed.

Materials needed

a stopwatch   a surveyor's tape   chalk

Procedure
1. Work with three other students in the school yard. On the ground, mark your starting point with chalk.
2. Measure off 100 meters. Make another chalk mark there.
3. Have someone hold the stopwatch and act as timekeeper. The timekeeper will say "Go" and start the watch. When you have walked the 100 meters, the timekeeper will stop the watch and record the time.
4. Repeat the activity with each person walking the distance and, then, running the distance. Record these times too.
5. Figure each rate by dividing 100 by the number of seconds it took to go the 100 meters. The answer will be in meters/second.

Observations
1. What was your rate walking?
2. What was your rate running?

21

Matter

In this chapter, you will find out that

- Matter is anything that has volume and mass.
- Solids, liquids, and gases are the three states of matter.
- There are more than 100 different elements.
- Elements can be combined to form compounds.

Before you begin

These paints have been mixed to get the right colors. The paint will be applied with brushes. The paint cans and the paper are solid objects. The paints are liquids. When they dry, they will be solids and stick to the paper. You can smell the paint as it dries. The paint gives off an odor which travels through the air. The air is a mixture of gases.

Solids, liquids, and gases are all matter. What is matter? What are the properties of matter? The students mixed different compounds to get different color paints. What are compounds? How are compounds formed? Sarah has to climb the stepladder to get to the top of the mural. The ladder is made of metal. How is something made of metal different from something that is not made of metal? You will be able to answer all of these questions after reading this chapter.

What is matter?

You are probably in your classroom. Are most of the seats taken? Do you and your classmates have weight? If you answered yes to these questions, then you and your classmates are a kind of matter. Your body also has **mass**. Mass is the amount of **matter** in an object. The desks and windows in your classroom are matter. So is the air you breathe.

Do you know the difference between weight and mass? Let's take an imaginary trip into space. What will happen when you leave the pull of Earth's gravity? You will become weightless. You will float around in the cabin. 1   The weight of an object is a measurement of how much gravity is pulling on the object. When there is no gravity, there is no weight. Your body will still look the same and take up the same amount of space. That is because your body has mass. The mass of an object does not change in space.

On the moon, you would weigh ⅙ of your weight on Earth. The pull of gravity on the moon is ⅙ that of the Earth. But your mass would not change. You would still be made of matter.

The amount of space that matter takes up is called its **volume** (vŏl′yo͞om). Look at your textbook. With a metric ruler, you can measure its width, length, and thickness. 2   By multiplying these three measurements, you can find the volume of the book. The volume is measured in cubic centimeters.

What other objects in your classroom have volume and mass? Look at the lights. The glass of the light bulbs is matter. But light is not matter. Light does not have mass or volume. Light is a form of

mass:
The amount of matter in an object.

matter:
Anything that takes up space and has mass.

1

volume:
The amount of space an object takes up.

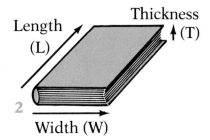

Length (L)   Thickness (T)

2

Width (W)

L × W × T = VOLUME (cm³)

energy. What are some other things in your classroom that are not matter? If you said heat or sound, you're right.

Three states of matter

Matter exists in three states. Your desk is an example of one state of matter. It is a **solid (sŏl'ĭd)**. A solid has a definite shape and volume. **Liquids (lĭk'wĭdz)** are another state of matter. Liquids, such as milk, flow freely and take the shape of their containers. They have definite volume. **Gases**, like the air we breathe, have no definite shape or volume. For example, if Bill opens a bottle containing ammonia gas in the front of the room, the gas will spread out to fill the whole volume of the room.

All matter is made up of tiny particles called **molecules (mŏl'ə kyoolz')**. The molecules in matter are constantly moving. The molecules in a solid move very slowly. They are close together and have very little space between them. So solids keep their shapes. 3 In a liquid, there is more space between the molecules. The molecules can move more than in a solid. So a liquid flows and takes the shape of its container. 4 The molecules in a gas are very far apart. 5 These molecules move very rapidly.

solid:
Substance that has a definite shape and volume.

liquid:
Substance that takes the shape of its container and has a definite volume.

gas:
Substance without a definite shape or volume.

molecule:
Smallest amount of a substance that has all the properties of the substance.

3   4   5

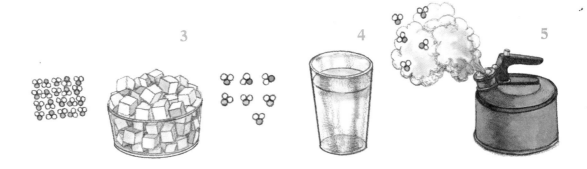

Activity

The movement of molecules

1. You will need a nonaerosol can of air freshener, a meter stick, and a watch with a second hand.

2. Measure the distance from the front of the room to your seat.

3. Your teacher will spray the air freshener at the front of the room. Note the time the spraying begins.

4. Raise your hand when you smell the spray. Record the time.

5. Make a class chart to show the results. Show the distances and the number of seconds. How long did it take for the molecules of spray to reach the back of the room?

Molecules in a gas are very far apart and move very rapidly. That is why you can smell the air freshener at the back of the room. The gas molecules spread out and fill the room.

✓ Check yourself

1. What is matter?
2. When does an object have mass but not weight?
3. What is volume?
4. What are three states of matter?

Elements

Molecules are made of small particles called **atoms**. Atoms are so tiny that the dot in a letter is made up of billions of them.

When a substance is made of only one kind of atom, it is called an **element** (ĕl′ə mənt). There are more than 100 elements. Most elements are found in nature. Some elements have been made in the laboratory by scientists.

Each element has a name and a chemical symbol. The symbol is a shorthand name for that element.

Eighty-nine elements are solids. The rest of the elements are liquids or gases. The solid elements are divided into metals and nonmetals. You can often tell metals from nonmetals by looking at them. Metals, such as iron and aluminum, have a shiny appearance. Nonmetals, such as carbon, have a dull appearance. The following chart shows some of the other differences between metals and nonmetals.

atom:
Smallest unit of an element.

element:
Substance made of only one kind of atom.

Element	Symbol
Hydrogen	H
Carbon	C
Oxygen	O
Chlorine	Cl
Mercury	Hg

333

Metals	Nonmetals
High melting points	Low melting points
Good conductors of electricity	Poor conductors of electricity
Will not let light through	Will let some light through
Can be pounded into sheets or drawn into wire	Will shatter when pounded
Example: aluminum 1	Example: sulfur 2

1

2

Compounds

When two or more elements are chemically combined, a **compound** is formed. Compounds are new substances. They are different from the original elements that form them.

Thousands of different compounds can be formed using the 100 elements. Some compounds, such as salt, are made of only two different elements. 3 Other compounds are made of many different elements.

Water is a compound. One molecule of water is formed when two atoms of hydrogen combine

compound: Substance made of two or more different kinds of atoms.

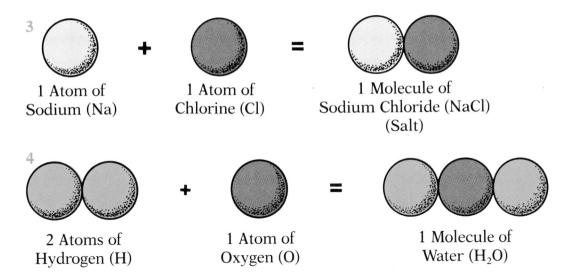

3
1 Atom of
Sodium (Na)
+
1 Atom of
Chlorine (Cl)
=
1 Molecule of
Sodium Chloride (NaCl)
(Salt)

4
2 Atoms of
Hydrogen (H)
+
1 Atom of
Oxygen (O)
=
1 Molecule of
Water (H$_2$O)

chemically with one atom of oxygen. **4** No matter what state of matter water is in, it is still water. Water is always made up of the same two kinds of atoms.

Activity

Classifying elements and compounds

1. You will need reference books on elements and compounds.
2. Copy the list of substances shown. **5**
3. If a substance is an element, write element next to it.
4. List the elements that make up each compound.

Substances
carbon dioxide
iron
oxygen
hydrogen
sodium chloride
carbon
water
aluminum
sulfur
sulfur dioxide

5

✓ Check yourself

1. What is an element?
2. About how many elements are there?
3. Name one way that metals and nonmetals are different.
4. What is a compound?
5. Is water a compound? What elements make up water?

Science in our world

What do plastic sandwich wrap, gasoline, and a lipstick have in common? They can all come from the same source—petroleum. Petroleum is an oily mixture of compounds that comes from under the ground or beneath the sea. It can be separated into more than 60 compounds. The compounds can then be put to many uses.

There are many compounds in our daily lives. The stainless steel fork you use to eat with is made of iron and carbon compounds. And probably the food at the end of your fork is also made of compounds.

Careers

Physical chemists are scientists who study how atoms and molecules form different compounds. Physical chemists usually work in laboratories.

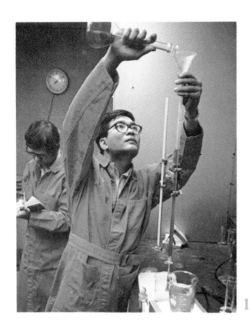

They test and examine compounds in all the states of matter. By studying the molecules in the air, physical chemists can help explain what causes our weather. They can also tell us a lot about the moon by examining rocks from the moon. Physical chemists are college graduates who often have advanced degrees. They may teach at a university. They may work for industry or for the government.

Review the main ideas

Matter is anything that has mass and volume. Volume is a measurement of the amount of space matter takes up. Matter is found in the solid, liquid, or gas state. All matter is made of molecules. When the molecules in a substance are made of only one kind of atom, that substance is an element. There are over 100 elements. The atoms of different elements may be combined to form the molecules of compounds.

Chapter review

Science vocabulary

Choose the items in Column B that best match the terms in column A.

A

1. matter
2. volume
3. weight
4. element
5. compound
6. atom
7. molecule
8. chemical symbol

B

a. substance made of two or more kinds of atoms

b. smallest amount of a substance that has all the properties of the substance

c. substance made of only one kind of atom

d. smallest unit of an element

e. shorthand name for an element

f. anything that has mass and takes up space

g. measure of the pull of gravity

h. measure of space taken up by matter

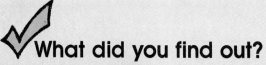

What did you find out?

Answer the following.

1. Solids, liquids, and gases are the three states of _____ .
2. More than _____ elements have been discovered.
3. Molecules move most rapidly in _____ .
4. Molecules of a compound are made up of two or more different kinds of _____ .
5. The 89 solid elements are divided up into _____ and _____ .

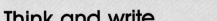

Think and write

1. When could an astronaut weigh less than his or her Earth weight?
2. You are standing in the back of a room. A container of carbon (a solid) is opened in front of a room. A container of hydrogen sulfide (a gas) is opened at the same time. Will you smell the carbon? Will you smell the hydrogen sulfide? Explain your answers.

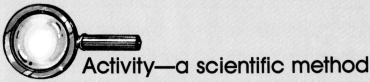 ## Activity—a scientific method

Purpose: to prove that the volume of a liquid does not change as shape changes.

Materials needed

a quart milk carton   a quart measuring cup

a shallow dish that holds one quart   water

Procedure

1. Fill the quart measuring cup with water.
2. Pour the water into the milk carton.
3. Pour the water from the milk carton into the dish.
4. Pour the water back into the measuring cup and measure it.

Observations

1. Did the shape of the water change?
2. In which container did the water take up the greatest volume? (careful!)

22

Matter changes

In this chapter, you will find out that

▢ We describe a substance according to its properties.

▢ There are two kinds of changes in matter: chemical and physical.

▢ New substances form in chemical changes but not in physical changes.

▢ Rusting is a chemical change in which iron combines with oxygen to form iron oxide.

Before you begin

Have you ever noticed how things change? Ice can become water, and water can become ice. Flour can be combined with other things to become a cake. Can you remove things from a cake and turn them back into flour and the other things?

Jane has a new bicycle. She is comparing it with her friend Ted's old one. There is no rust anywhere on Jane's bike. But Ted's has a rusty place on it.

Jane does not want any rust on her bike. She looks carefully at Ted's. She sees that it is scratched where the rusty place is. Then Ted shows her another scratch that is shiny, not rusty. He tells Jane the scratch just happened today. Why is the new scratch shiny and the old one rusty? If Jane's bike gets scratched, what can she do to prevent rust?

Have you ever seen gold rust? How about wood? Paper? Why do some things rust while others do not? What is rust?

Rusting

A bicycle is made of steel. Steel is a metal that is made up mainly of iron. Iron is a hard, gray, shiny substance. These are some of its **properties** (**prŏp′ər tēz**), or ways in which we describe iron. One of the substances in the air is **oxygen** (**ŏk′sĭ jən**). It has the property of being a gas. You cannot taste, see, or smell oxygen. So we say it also has the properties of being tasteless, colorless, odorless.

Rust is the substance that forms when iron comes together, or combines, with oxygen. Rust is a reddish-brown powdery, but solid, substance. These are its properties. Rust is not a metal, and it is not a gas. Its properties are different from iron or oxygen. You can see how rusting happens by using a piece of steel wool. Steel wool is a form of iron.

properties:
Traits that help to tell one thing from another.

oxygen:
A colorless, odorless, tasteless gas that makes up about one-fifth of air.

Activity

Making rust

1. You need a clean olive jar, a ruler, steel wool, and a small plastic cup.

2. Wet the steel wool. Then wedge it into the bottom of the jar. Set the jar upside down in a cup of water for several hours.

3. What happens to the steel wool?

4. Measure the height of the inside of the jar and then the height of the water in the jar.

5. Why does water enter the jar? How far into the jar does the water rise?

About 20 percent or one-fifth, of the air in the jar was oxygen. All the oxygen in the jar combined with the iron in the steel wool to make rust. When no oxygen was left, the rusting stopped. The water in the cup took the place of the oxygen. That is, as the oxygen was used up, the water moved up into the jar. The water filled about one-fifth of the space in the jar.

The air that is left in the jar is mostly **nitrogen** (**nī**′trə jən). Like oxygen, nitrogen is an invisible gas found in the air. But it does not combine with iron the way oxygen does. In fact, no other substance on Earth combines with iron the way oxygen does. Iron rusts only when it combines with oxygen.

nitrogen:
An odorless, colorless, tasteless gas. It makes up about four-fifths of the air.

Check yourself

1. List three properties of iron.
2. How much of the air is made up of oxygen?
3. What causes rust?
4. How much of the air is nitrogen?

Chemical changes

When iron combines with oxygen, a **chemical change** (kĕm′ĭ kəl chānj) takes place. In a chemical change, one or more new substances form. The new substance that forms when iron rusts is **iron oxide** (ī′ərn ŏk′sīd′). This is the chemical name for the substance we call rust. Iron oxide, or rust, has different properties from the iron and oxygen it is made up of. You have seen what some of these properties are. For example, iron is gray, oxygen is colorless, and rust is reddish-brown. What are some other differences in properties between these substances?

Sometimes, water is needed for a chemical change to take place, even though the water is not part of the reaction. This is what happens with rusting. Water is needed to help the combining of oxygen with iron.

Many chemical changes cannot be easily **reversed** (rĭ vûrst′). That is, the new substances cannot be broken down to form the original substances. The rusting of iron is not easy to reverse. You cannot change rust back to iron and oxygen.

chemical change:
A change in which one or more new substances form.

iron oxide:
The substance formed when iron and oxygen combine; commonly known as rust.

reversed:
Undone.

344

Activity

Chemical change

1. Look at each picture.

2. Is a chemical change taking place in each picture?

3. What evidence do you have that a chemical change is, or is not, taking place?

Burning is a chemical change. When a piece of paper is burned, gases and carbon ash are formed. These cannot be changed back into paper.

When you eat food, chemical changes take place in your stomach and in your intestines. The food that you eat is changed into substances your cells can use.

Melting is not a chemical change. The water can be changed back into ice.

Slicing a tomato is not a chemical change. The tomato may be cut into pieces, but it is still a tomato.

Can you think of any other examples of chemical changes? Remember, for a change to be chemical, a new substance must be formed.

Physical changes

Melting ice and slicing tomatoes are examples of **physical changes (fĭz′ĭ kəl chānjĕz)** in matter. When a physical change takes place, no new substance is formed. Ice and water are two forms of the same substance. So changing one to the other is a physical change.

Many physical changes are fairly easy to reverse. That is, it is possible to get back the original substance. Is the change between water and ice one of these?

When you slice a tomato, the tomato has a different shape. 1 But it tastes the same. And all its other properties are the same. It still has the same substances in it that a whole tomato has.

Suppose you mix sugar with water until the sugar disappears, or dissolves. Is this a physical change? Explain your reasoning.

What other examples of physical changes can you think of? Explain why your examples are physical and not chemical changes.

physical change:
A change which is usually fairly easy to reverse and in which no new substance is formed.

346

✓ Check yourself

1. What kind of change occurs in rusting?
2. What is another name for rust?
3. What is formed in a chemical change?
4. Describe a chemical change that takes place in your body.
5. What kind of change is the melting of ice?
6. Describe how physical changes and chemical changes are different from each other.

Science in our world

Rusting can damage things people use. This automobile shows the results of uncontrolled rust. 2

Coatings are used to prevent rust on iron and steel. Oils, paints, and other coatings seal out the air and water.

Look at a tin can. The can is mostly steel. It has two types of coatings, or sealers. 1 The coating on the inside of the can keeps food from reacting with the can. The coating on the outside protects the can from rusting. The outside coating of the can is tin. You can scratch the surface with a sharp nail or scissors. This exposes the steel underneath. The steel at the scratched place will rust. Paint is also often used to coat steel. What objects can you name that are painted to prevent rusting?

Some steel companies have developed special self-coating steels. These steels rust to form their own special coating. Such steel does not have to be painted. 2 This means a large saving to the building owner. The rusted surface protects the remaining steel. Ordinary rust does not form a protective coating on steel. It looks uncared for.

1

2

Careers

More than 500,000 people work in the chemical business in the United States. They make the raw materials used to make other materials. **Chemical workers** make many substances that are important to our lives. These include plastics, fertilizers, fabrics, dyes, and medicines. Chemical workers use machines and other modern equipment. 3 They keep records and test materials. Some of them repair equipment.

3

Chemical workers must be dependable and accurate. They must be able to follow directions.

Some chemical workers go to college after high school. Others go to trade school. They also learn from on-the-job training.

Review the main ideas

Iron can combine with oxygen to form iron oxide, or rust. Water helps in this process. This is a chemical change. In chemical changes, new substances are formed, and changes in properties take place. Another group of changes in matter are the physical changes. In physical changes, no new substances are formed.

Rusting is a chemical change that damages property. Rusting can be stopped in several ways. Iron can be coated or painted to keep out air and water.

Chapter review

Science vocabulary

Choose the items in column B that best match the terms in column A.

A

1. properties
2. iron oxide
3. physical change
4. chemical change
5. nitrogen
6. oxygen

B

a. gas that makes up one-fifth of the air
b. ways we describe substances
c. gas that does not combine with iron
d. change in which no new substances form
e. another name for rust
f. change in which new substances form

What did you find out?

Answer the following.

1. Steel wool is made mainly of the metal _____ .
2. Iron combines with oxygen to form _____ .
3. Two properties of iron are its _____ and _____ .
4. One difference between nitrogen and oxygen is that nitrogen (does/does not) combine with iron.
5. For iron to rust, both oxygen and _____ must be present.
6. Chemical changes usually (can/cannot) be reversed.
7. Melting of ice is an example of a _____ change in matter.
8. Slicing a tomato does not change most of its _____ .
9. To prevent rust, things made of steel may be _____ .
10. Some steels have a special _____ of rust.

Think and write

1. Does a chemical or physical change take place when leaves are burned? Explain your answer.

2. Does a chemical or physical change take place when a bottle breaks? 1, 2   Explain your answer.

3. Do other metals besides iron combine with the oxygen in the air? Investigate the following: aluminum, copper, silver, lead.

4. When you prepare foods by following a recipe, do chemical changes always take place? Can you think of any recipe in which the change is only a physical change? Explain your reasoning.

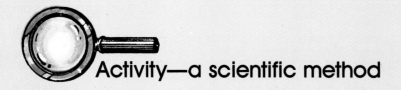

Activity—a scientific method

Purpose: to show the difference between a physical and a chemical change.

Materials needed

2 bottles
2 balloons
iron filings
measuring spoon
magnet
plastic cup
salt
water

Procedure

1. Fill one bottle half full of water.

2. Place your hand over the mouth of the bottle. Shake the bottle.

3. Take 1 level measuring spoon of iron filings.

4. Empty the water from the bottle. Immediately pour the iron filings into the bottle.

5. Place your hand over the mouth of the bottle. Shake the bottle. Some of the filings should stick to the sides.

6. Stretch the opening of a balloon. Use it to cap the bottle.

7. Measure out a spoon of iron filings. Pour those filings into the second (dry) bottle. Cap the bottle with a balloon.

8. Put these bottles aside until the next day.

9. Now, take a spoon of iron filings. Put it into a plastic cup. Mix in two spoons of salt.

10. Record what the mixture looks like.

Observations

1. How can you separate the mixture you made in step 9?

2. Try stirring the mixture with a magnet. What happens?

3. Keep stirring until you have gotten all the iron out of the mixture.

4. What kind of change was this?

The following questions refer to the bottles with the balloon caps. They are to be answered the next day.

5. What has happened to the filings in the wet bottle?

6. What has happened to the balloon on that bottle? Why?

7. What happened in the other bottle?

8. Was there a chemical change? How do you know?

23

Communicating today

In this chapter, you will find out that

- Radio and television waves travel through the air in different ways.
- Communications satellites speed communication.
- Communications satellites relay messages from one place on Earth to another.
- Computers store and process information.
- Computers can communicate with other computers.

Before you begin

About 500 years ago, Columbus came to America. He discovered many wonderful things. It took months for him to send word of his discoveries to Europe. He had to write the message by hand. Then he had to send that message by ship.

Even 100 years ago, it would still have taken 10 days to get a message across the ocean. Today, the same message will get there in seconds. In this chapter, you will learn about communications satellites (kə myoo′nĭ kā′shənz **sat**′l īts′). Together with other inventions, they have changed the way people communicate, or exchange information.

Do you know how communications satellites work? Do you know how satellites speed communications? Did you know that computers can communicate with one another?

Electronic communications

Think about all the communication instruments you use every day. You use the telephone to **communicate** (kə **myōō**′nə kāt′). Radio and television are also used for communicating.

Do you know when the first radio signal was sent through the air? It was in 1895. An Italian electrical engineer named Guglielmo Marconi sent it.

How does a radio work? A **radio transmitter** (trăns **mĭt**′ər) produces the radio waves. It also sends them into the air. The message is coded in a special way onto the waves. You know this as the AM and FM on a radio. 1 The waves go in all directions. Some radio waves go high into the air and bounce back. Others stay close to the ground. The waves are picked up by a **radio receiver** (rĭ **sē**′vər), or your radio.

Do you know when the first television sets were sold? It was 1939. The first real television program in the United States was seen in 1940. The early programs were only black and white. Not until 1953 did color television come into use.

Television waves travel in straight lines. They can only travel short distances. So television waves

1

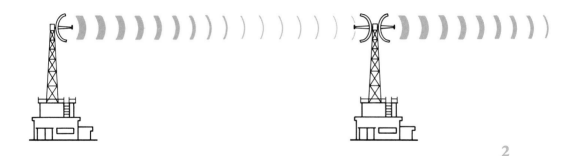

are relayed from tower to tower. **2** These towers are called relay towers. Each tower **amplifies** (ăm′plə fīz), or strengthens, the waves. Satellites are also used to relay television waves.

Today, **lasers** (lā′zərz) can also be used to send radio and television messages. Even telephone messages can be sent by laser. Lasers travel in one direction. But they travel far.

amplifies:
Strengthens.

laser:
A device that strengthens light into beams with the same wavelength.

Activity

Communications timeline

1. You need reference books on the discovery and development of communication, drawing paper, old magazines, scissors, and glue.

2. Start your timeline with cave drawings. Include scribes, the invention of the printing press, AM radio, FM radio. television, and the telephone on your timeline.

3. Cut out pictures from magazines and glue them on your timeline. How many different ways of communication are on your timeline?

357

Communications satellites

The problem of handling large numbers of telephone calls between continents was a difficult one. In 1965, the first step was taken to solve that problem. That was the year that Early Bird was launched.

Early Bird was a communications **satellite**. A satellite is a spacecraft that orbits, or circles, the Earth. A communications satellite serves a special purpose. It is used to get messages from one place on Earth to another. 1 Early Bird weighed 38 kilograms (about 17 pounds). It could handle 240 telephone conversations. Communications satellites now in orbit weigh 790 kilograms (359 pounds) and handle 6000 calls. The next satellites will be capable of relaying 12,000 calls at a time.

satellite:
A spacecraft that
orbits the Earth.

1

Communications satellites transmit many kinds of long-distance messages. They carry television programs from one part of the world to another. While sitting in front of your television set, you see and hear what is happening thousands of kilometers away. You see and hear it as it happens.

✓ Check yourself

1. How do radio waves travel?
2. How do television waves travel?
3. What happens to television waves at a relay tower?
4. What is the job of a communications satellite?

How satellites transmit messages

The United States has a communications satellite system. Messages can be sent all around the world. Signals are sent from transmitting stations on the Earth. The signals are beamed to a satellite by special antennas. **2** The satellite sends the

2

signals back to the Earth. The signals are picked up by receiving stations around the world.

Some signals are short, high-frequency waves called **microwaves** (mī′krō wāvz′). Microwaves travel above the Earth's surface. They grow weaker and weaker as they travel farther and farther from their source. 1 However, some communications satellites can amplify waves. Satellites that amplify waves are called **active satellites.** These satellites are powered by solar energy, or energy from the sun. Telestar was the first active satellite to send television messages between the United States and Europe. It was launched by the United States in 1962.

Other communications satellites do not amplify waves. These satellites are called **passive satellites.** Echo I was a passive satellite launched by the United States in 1960. Passive satellites just reflect the signals. Therefore, the signals are very weak. So special equipment on the Earth is used to pick up the signals. 1

microwaves:
Short, high frequency waves.

active satellite:
A satellite that amplifies waves.

passive satellite:
A satellite that just reflects signals.

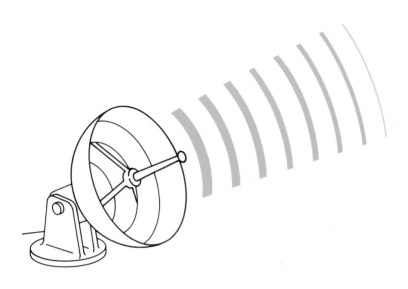

Check yourself

1. How are signals sent to satellites?
2. What are short, high frequency waves called?
3. What happens to microwaves as they travel farther and farther away from their source?
4. What powers some satellites?

Computers

Computers (kəm pyoo'tərz) have also improved communications. A computer is a machine that stores and handles information. It does this very, very quickly. A computer can find a piece of information thousands of times faster than a person can.

Every computer has at least four main parts. There are the memory and the central processing unit. Also, there are input and output devices. 3

computer:
A machine that stores and handles information.

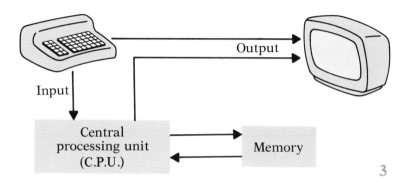

Input

Output

Central processing unit (C.P.U.)

Memory

3

361

Input devices are used to get information into the computer. People use input devices to tell the computer what to do. Output devices permit the computer to communicate with people. The output device shows what the computer has done. A video terminal (**vĭd′**ē ō′ **tûr′**mə nəl) is a common input/output device. It has a keyboard like a typewriter. That is for the input. It also has a screen like a television. That is the output. 1   A computer's output may take the form of numerals, words, graphs, or pictures. Some computers can even output voice messages and music.

A computer's memory stores information. All input is sent directly to the memory, where it is kept until needed. The main difference between a large computer and a small one is the size of its memory. A computer's memory stores more than information. It also stores instructions on how to use the information. Instructions and information are put into a computer as part of a program.

The computer program is what gets the machine

to work. Without a program, a computer would be less useful than a lightbulb. No computer can work without a program. Computer programs may be stored on magnetic discs or tapes. 2 Tape storage is one form of memory.

A computer may be programmed to play chess. It may also be programmed to perform blood tests in a hospital. With the right program, a computer may do bookkeeping or help you to make airline reservations.

2

The central processing unit is known to programmers as the CPU. It is in the CPU that all the work takes place. The CPU is the computer's "brain." It makes all the computer's parts work together to process information. A part of the CPU is also able to do arithmetic and make logical decisions.

Let us see how the whole thing works. Suppose you work for an airline. You have a customer who wishes to fly from New York to Los Angeles. You use the computer keyboard to communicate with the computer. You ask it what flights are available.

The CPU receives your inquiry. It checks the memory to see what flights are available. The CPU then directs the output to print a list of available flights. That list appears on the video screen. 1 When your customer chooses a flight, you enter the choice through the keyboard. The CPU will note that there is one less seat on that flight. It then sends that information to memory to make sure that that seat is not sold a second time. The CPU also sends a message to output to let you know that your customer has a seat on that flight.

✓ Check yourself

1. List the four main parts of a computer.
2. What is the job of a computer keyboard?
3. What is the main difference between a small and large computer?
4. What is a program and why is a program so important?

Computers and printing

Computers are changing the printing industry. They are doing work previously done by people. They do this work faster and more accurately than people can. They also cost less to run than people would earn for doing the same job.

Computers have caused many people in the printing industry to lose their jobs. At the same time, they have opened many new jobs in the computer industry. 2

2

365

Here is what happens at a newspaper that does not use computers: Reporters type their news stories on paper. These stories are given to people called typesetters. Typesetters put together steel letters to make sentences, paragraphs, and pages. Those letters will then be used to mold a printing plate. 1 Such a plate was used to print this page.

1

At a computerized newspaper, reporters type their stories directly into the computer. The computer then controls a platemaker. The computer arranges the stories into pages and drives the platemaker to make the printing plates.

One computerized newspaper has printing plants in several cities. Its main office is in New York City. First, the computer gets all the stories arranged into pages. Then the New York City computer communicates with computers in the other cities. This communicating may be done by telephone lines or by satellites. Then the computers in all the plants drive printing presses. Presses in all the plants print the newspaper at the same time. That means that the morning papers can be on the newsstands in all parts of the country at the same time. Before computers, people in California had to wait for their papers to be flown by plane from New York. That took many hours and cost a lot more money.

Home computers

Some people have small computers in their homes. 2 They find many uses for their computers. Home computers can be used to play games. They can be used to keep records of birthdays, addresses, and recipes. Home computers can be used as tutors. They can help people learn spelling, arithmetic, and other subjects. They also control home security systems and appliances.

In the future, almost every family will have access to a computer. Each family may have terminals that connect them to giant computers. The giant computers will be like today's encyclopedias. But they will contain a lot more information. That information will never become out of date. That is because it will be updated constantly. Just type a question on your keyboard. Almost immediately the answer will appear on the video screen.

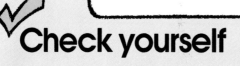

Check yourself

1. How have computers changed the way newspapers are printed?

2. How are stories typed at a computerized newspaper?

3. How do computers communicate with each other?

Science in our world

Written language made it possible to store ideas and pass them on from one generation to the next. Today, we have stored up a huge amount of information from the past.

Activity

How do libraries store information?

1. Go to the library and look at its resources.
2. List all the ways that the library stores information.
3. Next to each item on your list, name something stored in that manner.
4. Which is the most common way that the library stores information?
5. Which is the newest storage method?

Scientists are both withdrawers and depositors. They draw on the ideas of others. They also add their own ideas and discoveries. When you read about scientists' discoveries, you are a withdrawer. People who make new discoveries, are depositors in the bank of knowledge.

Videotapes, films, records, and computers also store information. How do computers compare with other storage methods? Can you think of any advantages books may have over recordings as a source of information? Do recordings or film have advantages that books do not?

Our huge bank of information makes us all richer. What if each generation tried to teach the next generation everything it knew without writing it down or otherwise recording it? Much information would be lost. Many things we have today would not exist.

The television set is just one example. One lifetime is not enough to collect all the scientific knowledge that makes television possible. Each generation gets a richer bank of information than the generation before it. Every generation adds to the bank. What new marvels do you think your generation will add to the bank? Think of some inventions that might make your children's lives richer and easier.

Careers

Computer programmers write computer programs. 1 The programs cause the computers to perform certain functions. Computer programmers find jobs in government, education, publishing, and all industries. More openings for computer programmers are created each year. This is an occupation with a growing future.

1

Computer programmers usually have college degrees. Some, however, are self-taught. Programmers must be good in mathematics. They must be logical thinkers and enjoy challenges. They are very well paid.

Review the main ideas

Throughout history, scientists and engineers have developed ways to improve communication. Printing presses, communications satellites, and computers improve communication. They bring information to more people. They make communication faster and cheaper.

Communications satellites receive, amplify, and transmit electrical signals that have been changed into microwaves.

Computers store and handle information. They have four main parts: memory, central processing unit, and input and output devices.

Our bank of information is continually growing. Each generation uses information gathered by previous generations. Each generation also adds to the bank of information.

Chapter review

Science vocabulary

Choose the items in column B that best match the terms in column A.

A	B
1. communicate	a. store information
2. satellites	b. exchange information
3. microwaves	c. objects that orbit Earth
4. amplify	d. strengthen
5. computers	e. short, high-frequency waves

✓ What did you find out?

Answer the following.

1. Microwaves grow (weaker/stronger) as they travel farther and farther from their source.

2. To be useful, a computer must have a _____ .

3. Information is stored in a computer's _____ .

4. Television messages are made up of _____ .

5. Three uses for home computers are _____ , _____ , and _____ .

Think and write

1. Describe how a satellite is used to send the audio, or sound, part of a television program.

2. Describe how you would use a home computer.

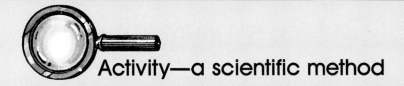 Activity—a scientific method

Purpose: to make a flowchart.

Materials needed

pencil   paper

Background

Computer programmers use flowcharts to design programs. Flowcharts contain three types of shapes. Circles mark the beginning and end. Rectangles hold "statements," or things to be done. Diamonds call for decisions. See if you can follow this flowchart.

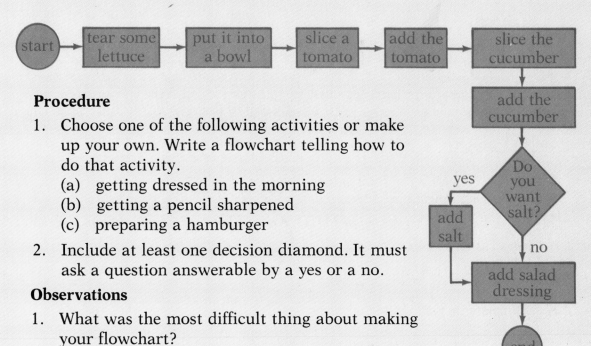

Procedure

1. Choose one of the following activities or make up your own. Write a flowchart telling how to do that activity.
   (a) getting dressed in the morning
   (b) getting a pencil sharpened
   (c) preparing a hamburger

2. Include at least one decision diamond. It must ask a question answerable by a yes or a no.

Observations

1. What was the most difficult thing about making your flowchart?

2. Do you think a computer could follow your flowchart? Why?

Glossary

abdomen (*ăb' də mən*) n. The end body part of an insect.

adapted (*ə dăpt'ĭd*) adj. Suited to an environment.

additive (*ăd'ĭ tĭv*) n. Something that does not occur in a food naturally.

adult (*ə dŭlt'*) n. A fully formed insect.

air mass (*âr măs*) n. A large quantity of the same type of air.

amplify (*ăm'plə fī'*) v. To strengthen.

antennae (*ăn tĕn' ē*) n. Parts on an insect's head used for taste, smell, and touch.

antibiotics (*ăn' tē bī ŏt'ĭks*) n. A group of wonder drugs that fight certain microorganisms causing disease in people.

artificial (*är' tə fĭsh'əl*) adj. Not found in nature; made by people.

asteroid (*ăs'tə roid'*) n. Tiny bits of planetlike material that orbit the sun.

astronomy (*ə strŏn'ə mē*) n. The study of heavenly bodies.

atoms (*ăt'əmz*) n. Tiny particles that make up all matter.

attract (*ə trăkt'*) v. To pull toward each other.

barometer (*bə rŏm'ĭ tər*) n. An instrument for measuring air pressure.

binoculars (*bə nŏk'yə lərz*) n. A hand-held optical device that can be looked through with both eyes.

blubber (*blŭb'ər*) n. Layer of oily fat in marine animals.

bone marrow (*bōn măr'ō*) n. The soft protein material found inside the bone.

calisthenics (*kăl'ĭs thĕn'ĭks*) n. A series of regular exercises.

calories (*kăl'ə rēz*) n. The amount of energy contained in foods.

canal (*kə năl'*) n. A passageway.

carbon dioxide (*kär'bən dī ŏk'sīd'*) n. A gas found in the air and used by plants during photosynthesis.

carnivore (*kär' nə vôr'*) n. Mammal that eats mostly meat.

cartilage (*kär'tl ĭj*) n. The firm but elastic material that makes up the nose and other parts of the body.

characteristics (*kăr' ĭk tə rĭs' tĭks*) n. The important or special features of a certain person or thing.

charge (*chärj*) n. The electrical property of particles of matter.

chemical change (*kĕm' ĭ kəl chānj*) n. A change in which one or more new substances form.

chlorophyll (*klôr'ə fĭl*) n. Material that gives green plants their color.

cirrus clouds (*sĭr'əs kloudz*) n. Thin, feathery clouds very high in the sky.

classify (*klăs'ə fī*) v. To put things in groups according to their likenesses and differences.

cochlea (*kŏk'lē ə*) n. The inner ear, containing a liquid.

cocoon (*kə kōon'*) n. A case made by some insects that protects the pupa.

colony (*kŏl'ə nē*) n. Community of insects.

comet (*kŏm'ĭt*) n. A body thought to be made up of gases, dust, and ice and having a "tail."

communicate (*kə myōo'nə kāt'*) v. To exchange information.

compound (*kŏm'pound'*) n. Substance made of two or more different kinds of atoms.

compound microscope *(kŏm'pound' mī'krə skōp')* n. An instrument that uses 2 or more lenses.

computer *(kəm pyōō'tər)* n. A machine that stores and handles information.

concave lens *(kŏn'kāv' lĕnz)* n. A lens that is thin in the middle and thick at the ends.

cone *(kōn)* n. Cell in the retina that detects color.

conservation *(kŏn'sər vā'shən)* n. Saving a resource.

consumer *(kən sōō'mər)* n. The person who buys and uses foods and other products.

contour *(kŏn'tōōr)* adj. Following the outline of the land.

convex lens *(kŏn' vĕks' lĕnz)* n. A lens that is thick in the middle and thin at the ends.

cornea *(kôr'nē ə)* n. The clear, protective covering in the front of the eyeball where light first enters the eye.

crop rotation *(crŏp rō tā'shən)* n. The practice of not planting the same crop in the same field year after year.

cumulus clouds *(kyōō'myə ləs kloudz)* n. Big, puffy clouds. They are lower than cirrus clouds.

current *(kûr'ənt)* n. Flow of electric charges.

data *(dā'tə)* n. Information.

discharge *(dĭs chärj')* v. To lose an electric charge.

electromagnet *(ĭ lĕk'trō măg'nĭt)* n. A magnet that works only when there is an electric current flowing.

element *(ĕl'ə mənt)* n. A substance made of only one kind of atom.

emotion *(ĭ mō'shən)* n. A strong feeling, such as anger, hate, or love.

endangered *(ĕn dān' jərd)* adj. In danger of disappearing from the Earth.

endurance *(ĕn dōōr'əns)* n. The ability to work for a long time without tiring.

erosion *(ĭ rō'zhən)* n. Carrying away of topsoil by wind and water.

eyepiece lens *(ī'pēs' lĕnz)* n. The lens of a telescope that is near the eye.

fauna *(fô'nə)* n. All the animal life in an area.

femur *(fē'mər)* n. The largest bone in the human body; the thighbone.

flora *(flôr'ə)* n. All the plant life in an area.

focus *(fō'kəs)* v. To make clear and sharp, as in a focused picture.

food chain *(fōōd chān)* n. A way to describe how energy, in the form of food, goes from one living thing to another.

fortify *(fôr'tə fī')* v. To make more healthful.

friction *(frĭk'shən)* n. Rubbing an object against another, producing heat and sometimes static electricity.

front *(frŭnt)* n. The line where one air mass meets a different air mass.

fuel rods *(fyōō'əl rŏdz)* n. The part of a nuclear power plant that contains the uranium.

galvanometer *(găl'və nŏm'ĭ tər)* n. An instrument that detects small electric currents.

gas *(găs)* n. Substance without a definite shape or volume.

gem *(jĕm)* n. A rare, valuable, and long-lasting mineral.

generator *(jĕn'ə rā'tər)* n. A machine that changes mechanical energy into electrical energy.

geologist *(jē ol'ə jĭst)* n. Scientist who studies the structure of the Earth.

geothermal *(jē' ō thûr'məl)* adj. The kind of heat that comes from within the Earth.

geyser (*gī′zər*) n. A hot spring that sometimes erupts.

granite (*grăn′ĭt*) n. A hard, long-lasting, light-colored stone with specks of black and pink.

gravity (*grăv′ĭ tē*) n. A force that tends to pull things toward each other.

hive (*hīv*) n. Home in which bees live.

igneous rock (*ĭg′ nē əs rŏk*) n. Rock formed from melted rock that cools.

invert (*ĭn vûrt′*) v. To turn upside down, or reverse in position.

invertebrate (*ĭn vûr′tə brĭt*) n. Animal without a backbone.

iris (*ī′rĭs*) n. Colored part of the eye that surrounds the pupil and controls amount of light entering the pupil.

iron oxide (*ī′ərn ŏk′sīd′*) n. The substance formed when iron and oxygen combine; commonly known as rust.

joint (*joint*) n. The place at which two or more bones meet.

kilometer (*kĭl′ə mē′tər*) n. 1000 meters. A unit of distance.

kingdom (*kĭng′dəm*) n. Largest group under which organisms are classified.

larva (*lär′və*) n. First stage of insect life.

laser (*lā′ zər*) n. A device that strengthens light into beams with the same wavelength.

lens (*lĕnz*) n. A transparent object that bends light.

ligament (*lĭg′ə mənt*) n. The elastic material that holds together the movable joints of the body.

limestone (*līm′stōn′*) n. A layered stone that contains fossils and may be white, gray, yellow, or brown in color.

liquid (*lĭk′wĭd*) n. Substance that takes the shape of its container and has a definite volume.

luster (*lŭs′ tər*) n. The way a mineral looks when it reflects light.

magma (*măg′ mə*) n. Hot, liquid rock beneath the Earth's surface.

magnetic field (*măg nĕt′ĭk fēld*) n. Lines of magnetic force surrounding a magnet.

mammal (*măm′əl*) n. An animal that has hair on its body; the female produces milk for its babies.

marble (*mär′bəl*) n. A kind of limestone that had been under great heat and pressure.

marine mammal (*mə rēn′ măm′əl*) n. Mammal that lives in the ocean.

mass (*măs*) n. The amount of matter in an object.

matter (*măt′ər*) n. Anything that takes up space and has mass.

metamorphic rock (*mĕt′ə môr′ fĭk rŏk*) n. Rock changed by pressure and high temperature.

meteroids (*mē′tē ə roidz′*) n. Stony particles that travel through space.

microorganism (*mī′krō ôr′gə nĭz′əm*) n. Tiny living thing that can be seen only with a microscope.

microscope (*mī′krə skōp′*) n. An instrument that allows one to see things that are too small to see with the naked eye.

microwaves (*mī′krō wāvz′*) n. High-frequency waves.

migration (*mī grā′shən*) n. Seasonal movement of birds to another place.

molecule (*mŏl′ə kyo͞ol′*) n. Smallest amount of a substance that has all the properties of the substance.

natural resource (*năch′ər əl rē sôrs′*) n. Material that is found in nature and is of value to humans.

negatively charged (*nĕg'ə tĭv lē chärjd*) adj. Having more negative charges than positive charges.

neutral (*noo'trəl*) adj. Having an equal number of positive and negative electric charges.

nitrogen (*nī'trə jən*) n. An odorless, colorless, tasteless gas. It makes up about four-fifths of the air.

noise (*noiz*) n. Unpleasant sounds.

nuclear energy (*noo'klē ər ĕn'ər jē*) n. Energy released when atoms are split.

nuclear reactor (*noo'klē ər rē ăk'tər*) n. The device in which nuclei are split to get nuclear energy.

nucleus (*noo'klē əs*) n. The central part of an atom.

nutrient (*noo'trē ənt*) n. The part of food the body needs to help it grow or work.

objective lens (*əb jĕk'tĭv lĕnz*) n. The lens of a telescope that is farthest from the eye.

optical illusion (*ŏp'tĭ kəl ĭ loo'zhən*) n. Something that appears to be what it really is not.

orbit (*ôr'bĭt*) n. The path an object takes around another object.

ore (*ôr*) n. A mineral deposit from which a valuable and usable amount of a material can be taken.

organism (*ôr'gə nĭz'əm*) n. A living thing.

ornithology (*ôr'nə thŏl'ə jē*) n. The study of birds.

oxygen (*ok'sĭ jən*) n. A colorless, odorless, and tasteless gas that makes up about one-fifth of the air and is needed for life.

penicillin (*pĕn'ĭ sĭl'ən*) n. An antibiotic.

pesticide (*pĕs'tĭ sīd'*) n. Chemical used to kill pests such as insects and rodents.

phases (*fāz'ez*) n. The changing shapes of the moon.

photosynthesis (*fō' tō sĭn'thə sĭs*) n. The way in which green plants make food.

photovoltaic cell (*fō' tō vŏl tā'ĭk sĕl*) n. A device that changes solar energy directly into electric energy.

physical change (*fĭz'ĭ kəl chānj*) n. An easily reversible change in which no new substance is formed.

pitch (*pĭch*) n. The highness or lowness of a sound.

planet (*plăn'ĭt*) n. A body in space. It has no light of its own.

pore (*pôr*) n. Tiny skin opening that releases sweat onto the skin's surface.

porous (*pôr'əs*) adj. Having tiny holes.

positively charged (*pŏs'ĭ tĭv lē chärjd*) adj. Having more positive charges than negative charges.

posture (*pŏs'chər*) n. The position in which the body stands.

precipitation (*prĭ sĭp'ĭ tā'shən*) n. Water that falls from the air to the Earth. It may fall as rain, snow, sleet, or hail.

preservative (*prĭ zûr'və tĭv*) n. An additive that keeps a food from spoiling quickly.

prey (*prā*) n. An animal hunted by another animal for food.

primary consumer (*prī'mĕr'ē kən soo' mər*) n. Animal that eats green plants (producers) directly.

primate (*prī' māt'*) n. Mammal with grasping hands, including monkeys, apes, and humans.

process (*prŏs'ĕs'*) v. To treat something so that it changes.

producer (*prə doo'sər*) n. Living thing that can make its own food.

properties (*prŏp'ər tēz*) n. Traits that help to tell one thing from another.

protein (*prō'tēn'*) n. Nutrient used by the body to build muscles, nerves, and other body parts.

protist (*prō'tĭst*) n. Microscopic organisms that are neither plants nor animals.

pulse (*pŭls*) n. The beating of certain blood vessels under the skin.

pupa (*pyōō'pə*) n. The inactive stage in the life of an insect.

pupil (*pyōō'pəl*) n. The dark-looking part in the center of the eye through which light passes.

quarry (*kwôr'ē*) n. The place from which rocks are removed.

radiation (*rā'dē ā'shən*) n. The releasing, or giving off, of energy.

radio receiver (*rā'dē ō' rĭ sē'vər*) n. A device that picks up radio waves.

radio transmitter (*rā'dē ō' trăns mĭt'ər*) n. A device that produces and sends radio waves.

reflecting telescope (*rĭ flĕkt'ĭng tĕl'ə skōp'*) n. A telescope that uses a mirror to collect light.

refracting telescope (*rĭ frăkt'ĭng tĕl' ə skōp'*) n. A telescope that uses lenses to collect light.

refraction (*rĭ frăk'shən*) n. The bending of light.

repel (*rĭ pĕl'*) v. To push away from each other.

retina (*rĕt'n ə*) n. The back part of the eye, on which an image forms.

reversed (*rĭ vûrst'*) adj. Undone.

rock cycle (*rŏk sī'kəl*) n. Changes in rocks from one kind to another kind in a continuous cycle.

rod (*rŏd*) n. Cell in the retina that detects dim light.

rodent (*rōd'ənt*) n. Small land mammal with long front teeth, including mice, rats, and squirrels.

rotate (*rō'tāt'*) v. To move around a center, or axis.

sandstone (*sănd'stōn'*) n. A layered stone that comes in different colors and textures.

satellite (*săt'l īt'*) n. An object that orbits, or revolves around, another object. A spacecraft that orbits the Earth.

secondary consumer (*sĕk'ən dĕr'ē kən sōō'mər*) n. Animal that eats other animals.

sedimentary rock (*sĕd'ə mĕn'tə rē rŏk*) n. Rock formed from sediments or pieces of other rock.

simple microscope (*sĭm'pəl mī'krə skōp'*) n. A hand magnifier using one lens.

skeleton (*skĕl'ĭ tən*) n. The body framework, consisting of 206 bones.

slate (*slāt*) n. A natural stone, usually made up of layers of clay pressed together over thousands of years.

social insect (*sō'shəl ĭn'sĕkt'*) n. Insect that lives in a society.

society (*sə sī'ĭ tē*) n. An organized group of living things.

solar battery (*sō'lər băt'ə rē*) n. Group of photovoltaic cells connected together.

solar collectors (*sō'lər kə lĕk'tərz*) n. Panels used to catch solar energy.

solar energy (*sō'lər ĕn'ər jē*) n. Energy from the sun. It reaches Earth as light and heat.

solar system (*sō'lər sĭs'təm*) n. A system that includes one or more suns, planets and their moons, and other bodies.

solid (*sŏl' ĭd*) n. Substance that has a definite shape and volume.

species (*spē'shēz*) n. Smallest group into which organisms are divided.

speed (*spēd*) n. A measure of how far something travels per unit of time.

speedometer (*spē dŏm'ĭ tər*) n. An instrument in a car that tells how fast the car is going.

spice (*spīs*) n. A plant substance added to food to preserve it or to flavor it.

spine (*spīn*) n. The column of bones in the back, also called the backbone.

static electricity *(stăt'ĭk ĭ lĕk trĭs'ĭ tē)* n. Electricity at rest.

stratus clouds *(strā'təs kloudz)* n. Low, dark, rain clouds.

streak *(strēk)* n. The color of the powdered form of a mineral.

surface water *(sûr'fəs wô'tər)* n. Water that can be seen above ground.

sweat gland *(swĕt glănd)* n. Small tube in the skin that produces salty fluid.

talon *(tăl'ən)* n. A long, sharp claw. Birds of prey have a talon on the end of each toe.

temperate *(tĕm'pər ĭt)* adj. Mild climate.

tendon *(tĕn'dən)* n. Narrow, strong cord that attaches muscle to bone.

terrace *(tĕr'əs)* n. A raised bank of earth with a flat top.

thorax *(thôr' ăks')* n. Middle body part of an insect.

topsoil *(tŏp'soil')* n. The fertile top layer of soil in which plants can grow.

verbal communication *(vûr'bəl kə myōō' nĭ ka' shən)* n. Communication using words.

vertebrae *(vûr'tə brē)* n. The little bones that make up the spine.

vertebrate *(vûr'tə brāt')* n. Animal with a backbone.

volume *(vŏl'yōōm)* n. The loudness or softness of a sound. The amount of space an object takes up.

warm-blooded *(wôrm'blŭd'əd)* adj. Having a body temperature that does not change with the outside temperature.

weathering *(wĕth'ər ĭng)* n. Breaking apart and breaking down of rocks on the surface of the Earth.

Pronunciation key

pat	ă	boot, fruit	o͞o
aid, fey, pay	ā	pop	p
air, care, wear	â	roar	r
father	ä	miss, sauce, see	s
bib	b	dish, ship	sh
church	ch	tight	t
deed	d	path, thin	th
pet, pleasure	ĕ	bathe, this	*th*
be, bee, easy, leisure	ē	cut, rough	ŭ
fast, fife, off, phase, rough	f	circle, firm, heard, term, turn, urge, word	û
gag	g	cave, valve, vine	v
hat	h	with	w
which	hw	yes	y
pit	ĭ	abuse, use	yo͞o
by, guy, pie	ī	rose, size, xylophone, zebra	z
dear, deer, fierce, mere	î	garage, pleasure, vision	zh
judge	j	about, silent, pencil, lemon, circus	ə
cow, out	ou	butter	*ər*
took	o͝o		

Index

adaptation to environment
  of birds, 49, 52, 65
  defined, 52
air
  and carbon dioxide, 20, 26, 29
  movement of, 196–197, 209
  oxygen in, 26, 29, 342
  travel of light through, 302, 313
air masses, 196, 209
  defined, 196
  warm and cold air masses, 196, 197,
    200, 205
air pressure, 201, 209
animals, 5, 10, 11, 15, 28, 29
  dependence on water, 179, 182
  and food chains, 49
ants, 33, 36–37, 42–43, 44, 45
  kinds of, 37
  life cycle of, 36–37
asteroids, 224, 227
atoms, 151, 213, 237–238, 239, 240, 337
  defined, 237, 333
  and nuclear energy, 237–238
  in water, 334–335
average speed, 317–318, 325
  finding average speed, 319–320
  straight-line characteristics of, 317, 321
  uses of in daily life, 323, 324

batteries, electrical, 261, 265, 272–273,
    276
  dangers of, 270
  kinds of, 269, 270
  and series circuits, 268
bees, 33, 40, 45
  beehives, 39, 40
  and making honey, 39–40
  queen bee and worker bees, 38, 39
  uses of, 45
Bell, Alexander Graham, 275–276
binoculars, 308–309, 310, 313
birds, 49, 65
  eyesight of, 55

general characteristics of, 50–52, 65
laws to protect, 64
migration of, 56
birds of prey, 49, 64
  hawk, 55–56
  as hunters, 53, 54, 55, 65
  lists of, 54, 56
  peregrine falcon, 53–54
  place in food chain, 59
  vanishing, 60–61, 65
bones, of human body, 85, 99
  bone marrow and cartilage, 88–89
  healing of broken bones, 97
  joints of, 90–91
  make up of, 87–88, 89
  number of, in body, 86
brain
  and hearing, 285, 291, 297
  and seeing images, 108, 113
breathing, 26–27, 29

Cade, Dr. Tom, 62
calories, 123–124, 117, 127
  calorie chart (diagram), 123
  counting calories, 124–125
  defined, 123
carbon dioxide, 19, 20
  defined, 26
  and green plants, 20
  and photosynthesis, 26–27
chemical changes, 341, 344–346, 349,
    352–353
chemical energy, 263, 264, 269, 276
chemical pollution, 49
  DDT, 59, 61–62, 65
  pesticides, 59
chlorophyll, 19, 20, 29
circuits. see electrical circuits
classification, 7
  defined, 4
  of elements and compounds, 335
  of organisms, 3
clouds, 181, 191, 195, 196, 209

formation of, 202, 211
kinds of, 202–203
comets, 224–225, 227
communication, 295–296, 297, 368–370,
    371
  by computer, 361–368
  by laser beam, 357
  by radio and television, 356–357
  by satellite, 355, 358, 359–360
compounds, 329, 334–335, 337
  classifying, 335
  defined, 334
  uses in daily life, 336
computers, 355, 371
  defined, 361
  effect on printing industry, 365–367
  home computers and uses, 367–368
  how a computer works, 363–364
  parts of, 361–362, 363, 364
  programs for, 362–363
conductors, of electricity, 249, 264–265,
    277
conservation
  defined, 167
  of energy, 264
  of soil, 167, 168–169
  of water, 188
conserving water, 179, 188
  how to (chart), 189
  need to, 190
consumers
  as buyers, 136
  food laws protecting, 140, 141, 142, 143
  primary and secondary consumers,
    57–58
convex lenses, 304–305, 313
crystals, in minerals, 151, 154, 155, 156
current, electrical, 267–268, 269, 276, 277
  defined, 265

distance, 317, 325
  how to measure, 318
  making distance-versus-time graph, 321

ear(s), 285, 295
Earth, 19, 33, 69, 157, 158, 165, 166, 167,
    179, 180, 181, 213, 214, 227, 231, 307,
    343
  and gravity of, 215, 330
  and the moon, 215–218

orbit and rotation of, 215
  as a planet, 220
  and the water cycle, 181
electrical charges, 247, 249, 250, 254
    255, 257, 265
  defined, 248, 265
  friction to produce, 250–251
  neutralizing, 252–253
electrical circuits, 261, 276
  galvonometer to detect, 272
  kinds of, 265, 267–268, 269
  parts of, 264
electrical generators, 271, 276, 277, 279
  defined, 264
  power sources for, 232–233, 236, 238
electricity, 231, 241, 247, 261, 276
  and circuits, 264–265, 267–269
  conductors and insulators, 265–266
  and magnetism, 272–273, 277
  sources of, 235–236, 264, 271, 272
  see also static electricity
electromagnets, 273
  defined, 274
  uses of, 274, 275–276
element(s), 161, 329, 333, 334, 337
  chemical symbols of, 333
  classifying, 335
  defined, 150, 333
emotion and emotions, 117, 125, 127
  and communication, 292–293, 297
endurance, 118–119, 127
energy, 58, 231, 240, 241, 276
  calories as measuring unit, 123–124, 127
  contrasted with work, 262
  kinds and forms of, 237–238, 239, 240,
    261, 262–263
  light as form of, 20, 330–331
  plant sources of, 23, 25
  and rest and sleep, 119–120
  solar (sun) energy, 214, 215, 234–236
  sources of, 232–233, 236
environment, 62
  adaptation of birds to, 49, 52
  effect of elephant on, 80
erosion, 175
  defined, 167
  and farming methods, 170–171
  and plant cover, 170
exercise, 117, 118–119, 127, 129
  effect on work, 126
eyes, 103

color-detecting cells of, 108
how to judge distances, 115
parts of, 104, 106–107

Faraday, Michael, 276
farming methods, 165, 175
    contour farming, 170
    crop rotation, 172
    terrace farming, 170–171
Fermi, Enrico, 241
flowering plants, 33, 45
    and bees, 39, 45
food, 19
    and calories, 123–124
    as energy source, 58
    produced by green plants, 20, 22–23, 29
food additives, 131, 141, 137, 139–140
    defined, 133
    examples of, 133–134, 135, 136, 142–143
    harmful additives, 142
    laws concerning, 140
    nutrients (vitamins, iron, etc.), 138
food chains, 49, 57, 59
    defined, 58
    effect of DDT on (diagram), 61
food preservatives, 133–134
foods, 131, 132
    coloring of, 135–136
    nutrients added to, 180
    preserving, 133–135, 145
    processing of, 136–137
forest lands, 172–173, 175
    laws protecting, 174
fossil fuels, 231, 241
friction, 250–251, 257
    and static electricity, 252–253

galvanometer, 272–273, 277
gas and gases, 19, 329, 333, 337
    carbon dioxide, 26
    defined, 331
    how sound travels through, 281, 287
    hydrogen, 334–335
    molecules of, 331, 332
    nitrogen, 343
    oxygen, 26, 27, 29, 334–335, 342
    stars composed of, 214
geothermal heat, 232–233
Goodall, Jane, 80–81
gravity, 224
    defined, 215

in space, 330
and sun and Earth, 215
green plants, 19, 27, 29
    and food chain, 58, 59
    and oxygen production, 27
    as producers, 20, 22–23, 25–26, 29, 57
    and sun as energy sources, 264

Halley's comet, 225
hearing, 281, 297
    importance of, 295
    and parts of the ear, 285
heat, 231, 241
    as energy, 263
    kinds and sources of, 231, 232–233
    in rock formation, 156, 161
    and water evaporation, 181
humidity, 195, 209
    defined and explained, 200
hydrogen, combined with oxygen to
    form water, 334–335

igneous rocks, 149, 158, 161
    defined, 155
    examples of, 155, 156
    formed from magma, 155
images, 113
    focused in the eye, 106
    optical illusions, 109
    and telescopes, 305, 306
insect colonies, 33, 36, 38, 45
insects, 33, 38, 45
    characteristics of, 34–35
    how insects communicate, 42
    life cycle of, 37, 41
    see also ants; bees
insulators, of electricity, 264–265, 277
invertebrates, 45
    defined, 34
involuntary muscles, 92–93, 99
iris, of the eye, 104–105
iron, 341
    element of, 150
    an ore, 151
    and oxygen in rust formation, 243, 343,
        349
    properties of, 342
    used to fortify food, 138
iron oxide, 349
    defined, 344
    oxygen and iron to form, 341, 342

joints of the body, 99
  kinds of, 90–91
  ligaments to connect, 91
Jupiter, 220, 221, 224, 226

kingdoms, 3, 6, 15

larva, 45
  stage in life cycle of insects, 37, 41
lens(es), 301, 313
  in binoculars, 308–309
  to correct eyesight, 301, 304
  defined, 304
  of the eye, 106–107
  in focusing for far and near, 108
  kinds of, 112, 304, 305, 309, 311
  in microscopes, 310–311
  in telescopes, 305–307
  uses of, 304
ligaments, 85, 91, 99
light, 103, 113, 301, 313, 330–331
  and the eye and seeing, 104–105,
    106–107
  and minerals, 154
  refraction (bending) of, 302
  travel speed of, 286, 297, 301, 302
light energy, 20, 29
Linnaeus, Carolus, 8, 13
liquids, 25, 329, 333
  defined, 331
  how sound travels through, 281, 287
  molecules of, 331
Lubbock, Sir John, 42

magnetic field, 271, 272–273, 276, 277
magnetism, 261, 272–273, 277
magnets, 271
  and electricity, 272–273
  electromagnets, 274–275
mammals, 69, 71, 81
  characteristics of, 70–71
  examples and kinds of, 72, 73, 74, 76–
    77, 79
  main groups of, 72–74, 81
  see also "true" mammals
Marconi, Guglielo, 356
Mars, 220, 222, 224, 226
mass, 329, 330, 337
matter, 329, 337
  defined, 330
  and mass and volume, 330

states of, 331
mechanical energy, 262, 264
  and electrical generators, 271, 276
Mercury, 220, 222
metals, 333, 334 (chart)
metamorphic rocks, 149, 158, 161
  defined, and examples of, 157
meteorologists, 195, 205, 208, 209
microscopes, 301, 310–311, 313
  defined, 5
  uses of, 310, 311, 312
minerals, 149, 161
  atoms of, 151
  as "building blocks" of rocks, 155
  characteristics of, 150–151
  elements of, 150, 151
  examples of, 150, 151, 152, 154
  to fortify food, 138
  ores and gems, 151
  properties of, 152–154
mirrors, 313
  used in microscopes, 311
  used in telescopes, 306
Mohs, Frederick, 153
molecules, 337
  atoms of, 333
  defined, 331
  movement of, 332
moon, 213, 215–218, 227
  distance from Earth, 216
  in orbit around Earth, 216–217
  phases of, 217–218
  reflected light of, 217
  as satellite of Earth, 218
moons, 213, 227
  planets having moons, 222
Morse, Samuel, 275
muscles, 85, 99
  effect of exercise on, 118, 127
  functions of, 92, 93
  kinds of, 92–93

natural resource, 165, 166
negative electrical charges, 250, 252
Neptune, 220
nitrogen, 343
noise, 290
  defined, 293
  effect on hearing, 295
nonmetals, 333, 334, (chart)
nonverbal communication, 292–293

nuclear energy, 241
   controversy about, 240
   defined, 237
   diagram of nuclear plant, 238
   made from uranium, 237–238
nuclear reactor, 231, 238
nutrients, 19, 29, 143
   added to packaged foods, 138
   defined, 22, 138
   proteins, 26
   sugar and starch, 22–23

ocean water, 191
   removing salt from, 185
optical illusions, 103, 113
   defined, and kinds of, 109
   moving pictures as example of, 111
organisms, 3, 15
   defined, 5
   kingdoms of, 6
   naming of, 8, 13, 15
oxygen, 19, 26, 29, 341
   defined, 26, 342
   with hydrogen to form water, 334–335
   and iron in formation of rust, 342, 343, 349
   and photosynthesis, 26–27

packaged foods, 131, 132, 138, 141, 143
   additives and preservatives, 133, 137
   food coloring, 135–136
   labels on, 140
parallel circuits, 261, 269, 277
peregrine falcon, 53–54, 64, 65
   endangered, 60–61
   rescue program for, 62–63, 65
photosynthesis, 19, 22, 29
   and carbon dioxide, 26–27
   defined, 20
physical changes, 241, 346, 349, 352–353
   defined, and examples of, 346
physical fitness, 118, 126, 127
   and the emotions, 125
   importance of sleep and rest, 119–120
   posture, 122–123
pitch, of sound, 297
   defined, 289
planets, 213, 227
   defined, 219
   Earth, 219–220
   names of, 220, 221, 222

rings around, 221
   and the solar system, 224
   and their moons, 222
plants, 5, 10, 11, 15, 28, 31
   dependence on the sun, 214
   dependence on water, 179, 182
   effect on soil erosion, 170, 172–173
   as producers of oxygen, 26–27
   see also green plants
Pluto, 220, 222
positive electrical charges, 250, 252
posture, 122–123, 127
precipitation, 191, 209
   defined, 181
primates, 76–77
producers
   defined, 20
   of food, 22, 29
   and green plants, 57
protein(s), 29
   bone marrow as, 88
   defined, 26
   from plants, 26
protists, 5
pulse rate, 119, 127
   how exercise affects, 129
pupils, of the eyes, 103, 113
   defined, 104

queen ants, 36, 45
queen bees, 38, 45
quartz, 154

radiation energy, 239, 240, 241
radio and radio waves, 355, 356
reflected light
   of moon, 217, 227
   of the planets, 220
refraction of light, 313
   defined, 302
   see also binoculars; lenses; microscopes; telescopes
resistance, electrical, 264–265, 267, 269, 276, 277
rest, 117, 119–120
retina, 103, 112, 113
   defined, 106
   and focusing light (diagrams), 106, 107
rock cycle, 149, 158–159, 161
rocks, 149, 161, 163
   classification of, 155–157

grouping of, 149
  moon rocks, 226, 337
  uses of, 149, 160
  weathering of, 158
rust and rusting, 341, 342–343, 349
  chemical name (iron oxide), 344
  damaging effects of, 347
  prevention of, 348

Sagan, Carl, 227
salt, 133, 143, 185
satellites, 213, 371
  communication satellites, 355, 358, 371
  defined, 218, 358
  powered by solar energy, 360
  of the solar system, 218, 222
  and transmitting of messags, 359–360
Saturn, 220, 221, 222, 226
scientists, 3, 5, 6, 10, 13, 15, 54, 59,
    60, 61, 64, 65, 77, 81, 95, 142, 224,
    225, 227, 241, 301, 304, 305, 310,
    312, 313, 371
geologists, 155
  and information bank, 369–370
  and Latin language, 8
  physicists, 241
sedimentary rocks, 149, 158, 161
  defined, and examples of, 156
series circuits, 261, 267–268, 277
skeleton, human, 85, 99
  bones of, 87–89
  defined, 86
  functions and uses of, 87
skin, of human body, 85, 95–96, 99
sleep, 117, 119, 120, 127
societies, 33, 36
soil, 165, 175, 177
  composition and formation of, 166, 167
  erosion of, 167, 170–171
  and farming and grazing methods,
    170–171, 172–173
  as a natural resource, 166
  topsoil, 167, 175
  worn-out soil, 172–173
soil conservation, 165, 167, 168–169
  and Soil Conservation Service, 184
solar energy, 234–236, 241, 264
  to power communication satellites, 360
solar system, 213, 220, 224–225, 227
  composition of, 224
  defined, 224

solids, 25, 329, 333, 337
  defined, 331
  how sound travels through, 281, 287
  mineral characteristics of, 150, 151
  molecules of, 331
sound, 281, 285, 290, 297
  as communication, 291–293
  noise (unpleasant sound), 293
  qualities of (volume and pitch), 288–289
  rate and speed of, 286–287
sound waves, 283, 297
  and the ear, 285
  effected by distance, 284
  rate of travel, 287
species, 3, 13, 15
  defined, 7
  differences within, 9–10
speed, 317, 324
  defined, 318
  measuring, 318, 327
  see also average speed
starches, 23, 24, 29
stars, 213, 214, 227, 229
  composed of gases, 214
static electricity, 247, 257
  defined, 248
  and electrical charges, 248, 249, 250
  examples of, 255, 256
  and friction, 250–251, 252–253
  and moisture, 255, 257
  neutral state of, 252, 255
sugar, 19, 29
  from green plants, 22–23
  as a source of fats and oils, 25
  as a source of nutrients, 23, 25
sun, 213, 227
  composed of gases, 214
  Earth's rotation around, 215
  source of heat and light, 214, 215
surface water, 180, 191

telescopes, 301, 310, 313
  giant telescopes, 307
  reflecting telescopes, 306
  refracting telescopes, 305
  for viewing planets, 220, 224, 227
television, 356–357, 370
  by satellite transmission, 358
temperature, 195, 209
  of the air, 196
  thermometer to measure, 197

tendons, 85, 94, 99
    defined, 93
time, 317, 321, 325
    importance in daily life, 322
    and speed, 318
trash, 231, 241
    as source of energy, 233
"true" mammals, 74, 81
    examples of, 76, 77, 79
    ten main groups of, 76–77, 79
    see also mammals

underground water, 180, 191
uranium
    and heat radiation, 239
    used to make nuclear energy, 237–238
Uranus, 220, 221, 222

Venus, 220, 222, 226
verbal communication, 291–292
vertebrates, 81
    defined, 70
vibration, 281, 282
    and production of sound waves, 283
vision (seeing), 103
    dependence on light, 104, 106
    near and far vision, 107
    nearsightedness and farsightedness, 112
    optical illusion, 109, 111
    seeing color, 108
    seeing images, 106–107
    vision problems, 112, 113
volume, 329, 337
    defined, 288, 330
    of matter, 329, 337
    of sound, 297

water, 179, 191
    atoms of hydrogen and oxygen, 334–335

and green plants, 20, 29
purifying processes, 185–187
and refraction of light, 302, 303
and soil erosion, 172, 175
sources of, 180
travel of sound through, 287, 297
water cycle, 188, 191
    explanation and drawing of, 181
water pollution, 28
    sources of, 182
water supply, 179
    to cities, 184–185
    and farming, 167
weather, 195, 209
    air pressure and weather, 201
    effects of on daily life, 208
    fronts, 196, 197
    gathering data for, 204–205
    and humidity, 200
    maps of, 205–207
    movement of, 195
    and the sun, 215
weathering of rocks, 158–159, 161
weight, 117, 127
    contrasted with mass, 330
    effect of eating and exercise on, 123–124
weightlessness, 330
wind
    as energy source, 236, 241
    as mechanical energy, 263
    and soil erosion, 172, 175
    and the sun, 215
wires, electrical, 264, 265
    in electrical generators, 271
    and magnetic field, 273, 277
work, 276
    scientific meaning of, 262

Photo credits

1: Shelly Grossman/Woodfin Camp; 2: Leonard Lee Rue III/Monkmeyer; 4. Russ Kinne/Photo Researchers; 5tl: Runk/Schoenberger/Grant Heilman Photography, tr: Grant Heilman Photography, c: Runk/Schoenberger/Grant Heilman Photography; 8: Granger Collection; 9l: Leo J. Fiedler, r: Werner H. Muller/Peter Arnold, Inc.; 10: Carl Anderson/Design Photographers International; 11l: Leonard Lee Rue III/Monkmeyer, r: Philcarol/Monkmeyer; 12: Ken Karp/McGraw-Hill; 14t: David Frazie/Photo Researchers, b: Alec Duncan/Taurus; 18: Grant Heilman Photography; 21: Ken Karp/McGraw-Hill; 22l: J. Alex Langley/Design Photographers International; r: Grant Heilman Photography; 23l: Lester Sloan/Woodfin Camp, r: J. Alex Langley/Design Photographers International; 28t: Sheila Turner/Monkmeyer, b: Richard Choy/Peter Arnold, Inc.; 32: David Overcash/Bruce Coleman, Inc.; 34: Robert Mitchell/Animals, Animals; 35: Norman R. Thompson/Taurus Photo; 37t: Michael Edereggen/Peter Arnold, Inc., b: Robert Mitchell/Animals, Animals; 38, 39t: Hans Pfletschinger/Peter Arnold, Inc., b: Richard Parker/Design Photographers International; 44: Hans Pfletschinger/Peter Arnold, Inc.; 45: Tim Eagen/Woodfin Camp; 48: Shelly Grossman/Woodfin Camp; 51t: Stephen J. Krasemann/Peter Arnold, Inc., b: Phil Dotson/Design Photographers International; 53: Ron Willocks/Animals, Animals; 54: Anthony Mercieca/Photo Researchers; 55t: Leonard Lee Rue III/Photo Researchers, b: Sidney Bahrt/Photo Researchers; 56: C. Allan Morgan/Peter Arnold, Inc.; 60: Peregrine Funel; 65: John Running/Stock, Boston; 68: New York Zoological Society; 70l: Tom Bledsoe/Photo Researchers, r: Runk/Schoenberger/Grant Heilman Photography; 71t: H. Uible/Photo Researchers, c: Steven C..Kaufman/Peter Arnold, Inc., bl: Thomas D. W. Friedmann/Photo Researchers, br, 72: Tom McHugh/Photo Researchers, 73t: Stephen J. Krasemann/Peter Arnold, Inc., bl: Leonard Lee Rue III/Photo Researchers, br: A. M. Bailey/Photo Researchers; 74t: Alfred B. Thomas/Animals, Animals; b: Alix Coleman/Grant Heilman Photography; 76l: John Colwell/Grant Heilman Photography, r: Grant Heilman Photography; 77t: Tom McHugh/Photo Researchers, c: Jacques Jaugoux/Peter Arnold, Inc., b: Fred Bavendam/Peter Arnold, Inc.; 79t: Hans Pfletschinger/Peter Arnold, Inc., bl: Helen Cruickshank/Photo Researchers, br: Harold Hoffman/Photo Researchers; 81: Wide World Photos; 84: Focus on Sports; 88: John Lei/Omni-Photo Communications/McGraw-Hill; 90: Peter Vadnai/McGraw-Hill; 91t: Mimi Forsyth/Monkmeyer, b: Karen Halverson/Omni-Photo Communications; 98: Ron Karten/Omni-Photo Communications; 102, 104l: Ken Karp/McGraw-Hill, r: Design Photographers International; 110: Ken Karp/McGraw Hill; 111: Peter Vadnai/McGraw-Hill; 113: Rhoda Sidney/Monkmeyer 116: Focus on Sports; 118: Dan Porges/Peter Arnold, Inc.; 119: Karen Halverson/Omni-Photo Communications; 120: Ann Hagen Griffiths/Omni-Photo Communications; 122: John Lei/Omni-Photo Communications/McGraw-Hill; 124: Peter Vadnai/McGraw-Hill; 125: M. Fiedler; 126: John Lei/Omni-Photo Communications; 127: Jay Dorin/Omni-Photo Communications; 130: Ken Karp/McGraw-Hill; 132t: Libby Corlett/Design Photographers International, b: Sepp Seitz/Woodfin Camp; 133: J. Alex Langley/Design Photographers International; 134t: Ken Karp/McGraw-Hill, b: Bohdar Hrynewych/Stock, Boston; 136: Ken Karp/McGraw-Hill; 137l: John Lei/Omni-Photo Communications, r: Ken Karp/McGraw-Hill; 138: Ken Karp/McGraw-Hill; 139: John Lei/Omni-Photo Communications/McGraw-Hill; 142: Cary Wolinsky/Stock, Boston; 143: Ann Hagen Griffiths/Omni-Photo Communications; 146–147: John Blaustein/Woodfin Camp; 148: Michael Collier/Stock, Boston; 150: Vance Henry/Taurus; 151t: Manuel Rodriguez/McGraw-Hill, b: Malcomb S. Kirk/Peter Arnold, Inc.; 152t: Grant Heilman Photography, b: Barry L. Runk/Grant Heilman Photography; 153t: Bruce A. MacDonald/Earth Scenes; 154t: Grant Heilman Photography, b: Runk/Schoenberger/Grant Heilman Photography; 155: Carolina Biological Supply; 156t: Runk/Schoenberger/Grant Heilman Photography, b: Carolina Biological Supply; 157t: Grant Heilman Photography, b: Tom McHugh/Photo Researchers; 160: Mimi Forsyth/Monkmeyer; 164: Craig Aurness/Woodfin Camp; 166t: Hal McCusick/Design Photographers International, b: Dick Durrance/Woodfin Camp; 167: John Livingston/Photo Researchers; 168: Russ Kinne/Photo Researchers; 169, 170: Ken Karp/McGraw-Hill; 171l: Cary Wolinsky/Stock, Boston, r: Klaus D. Francke/Peter Arnold, Inc.; 172: Leo J. Fiedler; 175: Jack Parsons/Omni-Photo Communications; 178: J. Alex Langley/Design Photographers International; 180l: Omni-Photo Communications, r: Leonard Lee Rue III/Monkmeyer; 182: Paul Stephanus/Design Photographers International; 183l: Douglas Herlihy/Design Photographers International, r: James Karales/Design Photographers International; 184: Los Angeles Department of Water &

Power; **185:** Tom McHugh/Photo Researchers: **188:** Environmental Protection Agency; **190:** Steve Allen/Peter Arnold, Inc.; **191:** Ken Karp; **194:** Jack Ellena/Design Photographers International; **197:** John Lei/Omni-Photo Communications/McGraw-Hill; **200tl:** Charles E. Schmidt/Taurus Photo, **tr:** Stephen Paulson/Taurus Photo, **b:** Hal McCusick/Design Photographers International; **202t:** Michael Melford/Peter Arnold, Inc., **bl:** John Tiszler/Peter Arnold, Inc., **br:** Daniel Brody/Stock, Boston; **204:** Hilda Bijur/Monkmeyer; **209:** Ann Hagen Griffiths/Omni-Photo Communications; **212:** Jim Moy/Design Photographers International; **214t:** NASA, **b:** Jim Theologus/Monkmeyer; **216t:** J. Alex Langley/Design Photographers International, **b:** David Greenberg/Design Photographers International; **219–225:** NASA; **227:** Wide World Photos; **230:** Peter Arnold; **232:** George Dodge/Design Photographers International; **233:** Judy Anderson/Peter Arnold, Inc.; **234:** George Dodge/Design Photographers International; **236tl:** NASA, **tr:** Ysbrand Rogge/Photo Researchers, **b:** George Dodge/Design Photographers International; **238:** John Zoiner/Peter Arnold, Inc.; **239:** Jay Dorin/Omni-Photo Communications; **241:** United Press International, Inc.; **244–245:** George Dodge/Design Photographers International; **246:** Carter Hamilton/Design Photographers International; **248, 253:** Ken Karp/McGraw-Hill; **251:** Amber Guild, Ltd.; **255:** Peter Vadnai/McGraw-Hill; **256:** Freda Leinwand/McGraw-Hill; **257:** Chris Reeberg/Design Photographers International; **260:** Kenn Goldblatt/Design Photographers Guild; **262:** Ken Karp/McGraw-Hill; **263l:** Panuska/Design Photographers International, **tr:** Jack Parsons/Omni-Photo Communications, **br:** NASA; **266, 268:** John Lei/Omni-Photo Communications/McGraw-Hill; **270:** Union Carbide; **271:** John Colette/Stock, Boston; **274t:** John Lei/Omni-Photo Communications/McGraw-Hill; **276:** The Bettmann Archive, Inc.; **280:** Camera Hawaii/Design Photo Communications; **282, 283:** Ken Karp/McGraw-Hill; **284:** Wil Blanche/Design Photo Communications; **286:** Hal Yaeger/Design Photo Communications; **287, 288, 289, 292:** Ken Karp/McGraw-Hill; **291:** Robert Capece/McGraw-Hill; **293t:** John Lei/Omni-Photo Communications/McGraw-Hill, **b:** Stephen J. Krasemann/Peter Arnold, Inc.; **296:** Mimi Forsyth/Monkmeyer; **300:** Nikon; **302:** James Theologus/Monkmeyer; **303:** Henry Monroe/McGraw-Hill; **305:** Di Storia della Suiza Museo; **307:** Mimi Forsyth/Monkmeyer; **308:** T. W. Putney; **310:** Ken Karp/McGraw-Hill; **311:** Bausch & Lomb; **313:** Georg Gerster/Photo Researchers; **316:** Lewis Frank/Focus On Sports; **319–322:** Ken Karp/McGraw-Hill; **324:** Nathan Simon/Omni-Photo Communications; **325:** John Lei/Omni Photo Communications/McGraw-Hill; **328:** Ken Lax/McGraw-Hill; **330:** NASA; **334l:** Barry L. Runk/Grant Heilman Photography, **r:** Linda K. Moore/Rainbow; **337:** James H. Karales/Peter Arnold, Inc.; **340:** John Lei/Omni-Photo Communications/McGraw-Hill; **343, 345tl:** Ken Karp/McGraw-Hill, **tr:** John Dunigan/Design Photographers International, **bl:** William Hubbell/Woodfin Camp, **br:** Ken Karp/McGraw-Hill; **346:** William Hubbell/Woodfin Camp; **347:** Charles Anderson/Monkmeyer; **348l:** Peter Vadnai/McGraw-Hill, **r:** U.S. Steel; **349:** J. Alex Langley/Design Photographers International; **351:** John Lei/Omni-Photo Communications/McGraw-Hill; **354:** James Karales/Peter Arnold, Inc.; **356:** Granger Collection; **360:** Noel Brooks/Design Photographers International; **362:** Stock, Boston; **363:** Freda Leinwand/Monkmeyer; **364:** John Lei/Omni-Photo Communications/McGraw-Hill; **365:** John Zoiner/Peter Arnold, Inc.; **366:** Paul Kirouac/Monkmeyer; **367, 370:** John Lei/Omni-Photo Communications.

respe